ML

Books are to be returned on or before
the last date below.

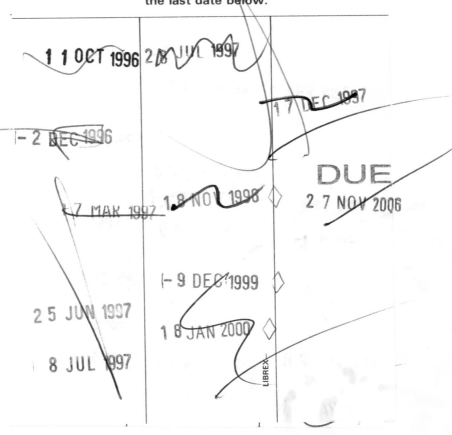

1 1 OCT 1996   2 8 JUL 1997

1 7 DEC 1997

-2 DEC 1996

7 MAR 1997   1 8 NOV 1998   DUE   2 7 NOV 2006

-9 DEC 1999

2 5 JUN 1997   1 8 JAN 2000

8 JUL 1997

LIBREX

AN ENGINEER'S GUIDE TO WATER TREATMENT

An Engineer's Guide to Water Treatment

George S. Solt
C.Eng. F.I.Chem.E. F.R.S.C.
Director, School of Water Sciences
Cranfield Institute of Technology

Chris B. Shirley
C.Eng. F.I.Mech.E. F.I.W.E.M.
President
Dewplan Group Ltd

Avebury Technical

Aldershot · Brookfield USA · Hong Kong · Singapore · Sydney

Published by
Avebury Technical
Academic Publishing Group
Gower House
Croft Road
Aldershot
Hants GU11 3HR
England

Gower Publishing Company
Old Post Road
Brookfield
Vermont 05036
USA

British Library Cataloguing in Publication Data
Solt, George S.
  An engineer's guide to water treatment.
  1. Water. Purification
  I. Title II. Shirley, Chris B.
  628.162

Library of Congress Cataloging-in-Publication Data
Solt, George S.
  An engineer's guide to water treatment/George S. Solt, Chris B. Shirley.
    p   cm.
  Includes index
  1. Water–Purification. I. Shirley, Chris. II. Title.
  TD430.S58 1991
  628.1'62–dc20

ISBN 1 85628 174 4

Printed in Great Britain at the University Press, Cambridge

Contents

List of figures   vii
List of tables   x
List of plates   xi
Foreword   xiii

Part 1 About water   1
    1   About water   3

Part 2 Pretreatment   9
    2   Pretreatment of surface waters   11
    3   Sedimentation   18
    4   Flotation   27
    5   The jar test   37

Part 3 Filtration of solids   47
    6   The background to design   49
    7   Filter designs   58

Part 4 Post treatment of potable waters   69
    8   Post treatment of potable waters   71

Part 5 Ion exchange and demineralization   77
    9   Ion exchange   79
    10   Weak and strong ion exchange resins   85

11	Base exchange (softening)	92
12	Dealkalization and weakly acidic resins	99
13	The principles of demineralization	107

Part 6   Ion exchange units and systems — 117

14	Counterflow	119
15	Degassing	135
16	Options in demineralization flow sheets	145
17	Mixed beds	152
18	A special use for ion exchange resin: nitrate removal	166

Part 7   Membrane processes — 175

19	Membrane processes in general and electrodialysis in particular	177
20	Reverse osmosis	186
21	Ultrafiltration and pretreatment for membrane processes	196

Part 8   Ultra-pure water — 203

22	General	205
23	Condensate polishing	213
24	Pyrogen-free (PF) water	233
25	Water for the electronics industry	241

Part 9   Deaeration — 249

26	The background to design	251
27	The practical design of heater deaerators	261

Part 10   Pollution and effluent treatment — 269

28	General	271
29	Biological filtration	283
30	Anaerobic digestion	293
31	Sewage treatment	306
32	Treatment of sewage sludge	321
33	Effluent problems in ion exchange	327
34	'Neutral effluent' — how neutral is it?	331

Appendix 1	Units of measurement in water analyses	339
Appendix 2	Glossary	347
Index		375

Figures

1.1	The water cycle	7
3.1	Schematic section of portion of lamellar sedimentation tank	20
3.2	Schematic section of hopper-bottom sedimentation tank	25
3.3	Schematic section of sludge recirculation tank	26
4.1	Dissolved air flotation tank	32
4.2	Flotation using static head	34
5.1	Relationship of pH to the alkalinity/free CO_2 ratio	41
7.1	Vertical pressure filter	60
7.2	Horizontal pressure filter	60
7.3	Rapid gravity filter	61
7.4	Integral backwash storage filter	61
7.5	Typical upflow filter design	66
8.1	Percentage HOCl in hypochlorous acid as pH varies	73
8.2	Chlorine breakpoint dosing	73
8.3	Basic chlorination control scheme	74
11.1	Resin affinity for calcium ions	94
11.2	Resin regeneration by sodium ions	94
11.3	Progress of ion exchange zone	96
11.4	Actual concentrations of ions within ion exchange zone	97
12.1	Reactions of CO_2 and water	99
12.2	Alkalinity, pH and free CO_2	101
12.3	Three methods of obtaining a soft water of zero alkalinity	102
13.1	Maximum permissible boiler drum contents	108
13.2	Classical demineralization flow sheet	115

14.1	Typical unit (co-flow)	120
14.2	Typical unit internals	120
14.3	Nozzle plate unit design	121
14.4	Typical liquid and air hold down designs	123
14.5	Typical regenerant collector systems	125
14.6	The PSB system	129
15.1	Concentrations of CO_2 in air/water	139
15.2	Countercurrent mass balance	140
17.1	Typical mixed-bed unit operation	153
17.2	Effect of leakage on pH and conductivity	159
17.3	Exhaustion of mixed-bed unit with low anion resin component	160
17.4	Exhaustion of mixed-bed unit with low cation resin component	161
17.5	Use of inert resin in mixed-bed unit	165
18.1	Development of chromatographic banding in exhaustion cycle	169
18.2	Nitrate breakthrough with conventional ion exchange resin	170
19.1	Electrodialysis and reverse osmosis membranes and concentration phenomena	180
20.1	Spiral wound module	191
20.2	Hollow fibre module	192
20.3	Three-stage RO system with concentrate recovery	193
22.1	Contaminant size spectrum	207
22.2	Variation of conductivity of pure water with temperature	211
23.1	Schematic diagram of typical HP boiler/turbine system	216
23.2	Schematic diagram of typical condensate polishing plant with loop recycle and external regeneration	232
25.1	Typical flow diagram for electronics industry	244
26.1	Heater deaerator	254
26.2	Concentration of O_2 in water/vapour	256
27.1	Heater box	265
27.2	Spray-and-tray deaerator	265
27.3	Scrubber design	267
27.4	Stork-type deaerator	267
29.1	Typical biofilter treatment system	286
30.1	Waste conversion	294
30.2	Metabolic pathways of methane formation	295
30.3	Process designs for anaerobic digestion	296
	a Conventional reactor	
	b High rate digester	
	c Contact process	
	d Anaerobic filter	
	e UASB process	
	f Fluidized-bed process	
31.1	Daily household water use	310

31.2	Typical large-scale domestic waste-water treatment plant	312
31.3	Typical activated-sludge process	315
31.4	Contact stabilization activated-sludge process	316
31.5	Application of treatment processes vs population	318
31.6	Septic tanks	320
32.1	Alternative routes for sludge disposal	322
A.1	Conductivity of ions vs concentration	345

Tables

9.1	Dissociation examples	81
9.2	The concentrations represented by different pH values	83
12.1	Comparison of treatment systems for softening a typical water	106
15.1	Henry's Law constant for CO_2	138
15.2	Packing data	143
16.1	Typical water analyses	146
16.2	Assumed chemical and water costs	146
16.3	Running costs resulting from selected process flow sheets	148
17.1	Units of measurement, conversion table	159
18.1	Quality of water denitrified by conventional ion exchange resin	168
26.1	Henry's Law constant for O_2	252
30.1	Methane production achieved from different substrates	300
31.1	Average (24h composite) analysis of crude sewage	307
31.2	Guidelines for the average flow of sewage from various establishments	311
31.3	Design of small treatment works: relative demands of power, land, capital and operator requirement	319
34.1	Typical demineralization plant chemical design	332
34.2	Typical components in effluent which affect the pH	335
A.1	Units of measurement: conversion table	343

Plates

Plate 1   Typical clarifier in industry   23
Plate 2   Raw-water clarification system at major power-station   24
Plate 3   Clarification plant showing solids recirculation clarifier
          with filters   25
Plate 4   Chemical dosing and regeneration station at major power
          plant   51
Plate 5   Pressure filter installation at Pembroke oil refinery   59
Plate 6   Typical municipal filter installations (inside view)   62
Plate 7   Typical municipal filter installations (outside view)   63
Plate 8   Integral backwash storage filtration plant and bulk
          chemical storage at nuclear site   64
Plate 9   Small packaged demineralizer comprising cation and anion
          vessels capable of producing quite high-quality water   110
Plate 10 Typical demineralization plant in a brewery   111
Plate 11 Major ion exchange plant at a nuclear power-station   112
Plate 12 Control panel for major ion exchange plant at a
          nuclear power-station   113
Plate 13 Typical demineralization plant at a major power-station   114
Plate 14 Typical pumping installation for membrane plant   188
Plate 15 Unusual photograph of a large reverse osmosis plant in an
          oil refinery   189
Plate 16 Prefabricated deaerator on its way to a major chemical
          complex   263
Plate 17 A complete water-treatment plant installed in Cuba   266

Plate 18   Ash-handling system under construction at a coal-fired
            power-station   274
Plate 19   Dairy effluent treatment plant under construction at
            Lord Rayleigh's dairy   275
Plate 20   Aerial photograph of industrial waste-treatment plant   281
Plate 21   Rotating biological contactor being installed at remote
            housing complex   308
Plate 22   Typical municipal sewage-treatment scheme including
            trickling biological filters   314
Plate 23   Sludge-settling system   324
Plate 24   An effluent pumping system at a major brewery   328

Foreword

In developed countries it is seldom that water can be used as abstracted without some form of treatment, be it for potable or industrial use. It is a long time since communities could take purity for granted or industries settle with a natural source of water suitable for their needs. The woollen industry of Yorkshire and the breweries at Burton-on-Trent are good examples of industries following water; now water can be made to suit industries requiring levels of purification previously unimaginable.

The biggest users of highly purified water are power generation (for boiler feed), pharmaceutical manufacture and the microchip industry, which needs the purest water of all. Reaching such purity is a specialized trade and depends on the quality of the raw water as well as that of the desired product. For example, the City of Glasgow advertises that it has the purest water supply of any industrial centre in the world, which is true. Unfortunately it also happens to be a very awkward starting material for reaching higher purities.

This book is designed to help non-specialists to understand the problems of water purification and is based on technical bulletins written for Dewplan Ltd by the authors, both engineers. Our experience was that water users, particularly in industry, were intimidated by the chemistry involved, and text books (such as they are) don't help much as they are mainly written by experts for other experts.

Actually, the amount of chemistry needed to understand our field is fairly modest. When we started writing our bulletins in 1971, our model reader was an old friend, Harold, who was a major boiler-house superintendent for a large chemical company. Harold was a well-qualified engineer, highly skilled in his profession, with water treatment as a peripheral but increasingly important interest. Thus Harold defined unknowingly the level of complexity, and our public liked it. We have now combined the collected output of almost twenty years into the book, and included work as yet unpublished.

Until recently our subjects were of separately identifiable interest to public water and industrial supplies. This is no longer true, and getting less so all the time. Water sources for public supply have deteriorated, and the EC directives now specify exact limits for a wide range of impurities in their product. More and more, public water undertakers are turning to processes previously used almost entirely by industry: ion exchange for denitrification and granulated activated carbon for the removal of organic impurities are two examples. Membrane and other new processes will serve both industrial and public supplies. We would certainly recommend that those professionals aspiring to treat or use water in either sector read the whole book in order to encompass the processes and practices detailed for both sectors.

We also wish to acknowledge the great help given in the writing of this book by David Moreau, master of many trades, for his unstinting support; Peter Jackson, Process Director of Dewplan for his invaluable technical opinions; John Hills, Director of the British Effluent and Water Association for permission to use references from BEWA publications; Bill Campbell and Ray Speight for their assistance with the section on pollution and, above all, Jan Green for coping with the secretarial load without complaint. Many others have assisted; all have our gratitude.

Chris Shirley
George Solt
February 1991

Part 1
ABOUT WATER

1 About water

Water is our most abundant raw material and, in its natural state, one of our purest. All the same, for many uses the available water is still not pure enough, and therefore some impurity or other often has to be treated or removed in order to make the available water fit to use. When we talk about the different properties of various waters, we are actually talking about the impurities which they contain. As a result, water treatment is a technology which is concerned with the impurities in waters, rather than with the water itself.

Water is a carrier

The simplest impurities in water are those which are merely carried in it, such as particles of sand. This suspended matter can be filtered out quite easily or (given time) a large part of it may settle out, though very fine or light particles may take an uneconomical time to do so.

Particles too small to be seen under a microscope are called colloids. These are so small that the constant movement of water molecules is enough to keep them floating and distributed through the liquid; they never settle out without assistance. Colloids are roughly in the size range 0.005 to 0.2 microns (a micron is 1/1000 mm). Such very fine particles cannot be filtered by conventional methods; any filter which retains them must be so fine that it raises a very high pressure loss and quickly tends to clog.

The lower size limit of colloids is an arbitrary thing, but we can consider any particles smaller than that to be in solution. In fact there are certain enormous molecules of organic matter occurring in natural waters which are quite as big

3

as colloids. These are sometimes called macro-molecules rather than colloids, but this is an exercise in name-calling from which we actually learn nothing.

Water is a solvent

Once we have passed through this rather uncertain size range, we come to materials which are truly in solution, like salts or detergents. They may be organic or inorganic, and they may be natural or come from some human activity. It is these which give different waters their main properties.

Apart from sea water and brackish water, there are basically three different kinds of waters:

Upland river waters   A glass of water drawn from the tap in Glasgow comes out with a yellowish tinge and a slight 'head' as though it was very thin beer. This is water which has run down the surface of hills over impervious rock, but through and over layers of peat, which is decaying vegetable matter. As a result it has picked up very little inorganic matter and it is soft but, on the other hand, it is full of dissolved and colloidal organic materials which have been picked up from the peat. These cause the yellowish tinge and the surface tension effect which makes the water foam. There are so many organic compounds in any one water that no one has ever analysed and isolated them all. As a group, they have a number of characteristics, some rather harmful, which we will consider later in the book. Because they are not clearly identifiable, the whole business of removing them from water calls for great skill. Wet, hilly, cool places such as Scotland, Wales and Canada tend to have waters like this.

Well waters   Waters which come from deep in the ground (and these include some spring waters) have percolated through mineral layers. These waters contain little organic matter, which the percolation tends to filter out. On the other hand, in passing through the underground minerals, the waters pick up inorganic salts and they tend to contain large quantities of calcium and magnesium, which makes them hard. Whereas upland river waters tend to be corrosive to steel, well waters tend to form scale in pipes and boilers.

Re-used water   A densely populated and industrialized country like the UK tends to re-use its waters. More and more of our supplies come from lowland rivers which contain the effluents from human activities upstream. The commonest of these is domestic sewage. The inorganic content of treated sewage is 50 to 100 ppm higher than that of the water supply to the township, so sewage not only adds a completely new class of organic matter to a river, it increases the contents of inorganic salts. Much of this additional inorganic matter is due to salts of sodium, which do not increase the hardness but do increase the total dissolved solids. There is also quite a lot of phosphate, which comes from domestic detergents.

The organic pollution from treated sewage will completely decompose, given enough time and dilution in the river. More and better sewage works will do much to reduce the organic content of the water by the time the next user receives it. On the other hand, sewage treatment does nothing to reduce the inorganic content and we can therefore expect the total dissolved solids in our raw waters to go on going up, as more water has to be re-used.

At the same time we already have to use many badly polluted waters; some industries actually use treated sewage directly, and some rivers are not much more attractive as water sources. It is in handling polluted waters that some of the most difficult treatment problems arise.

Water supports life

All waters are full of living organisms, spanning the size range from viruses to salmon. Except in the food industry, these are not often of great importance for industrial use but, where harmful, have to be neutralized for domestic consumption.

Water is cheap

Water's most outstanding property is its cheapness. No discussion of water should ever neglect this.

The most expensive drinking water now being widely distributed in the UK costs industry about 35p per m³. The worst quality water which is distributed in the UK contains little more than 500 ppm total dissolved solids. We might do better to think of these figures another way: water is a chemical which is being sold biologically safe, better than 99.95 per cent pure, at 35p per tonne (delivered).

Water treatment processes, too, are cheap. The most expensive process in water treatment is for industry and is total demineralization, which yields a product of less than 0.1 ppm total dissolved solids, which is 99.99999 per cent pure. The cost of this remarkable degree of purification, including amortization of plant, might be about another 50p per tonne.

No other process industry uses so cheap a raw material and produces a high purity chemical at so fantastically low a cost. It is this, rather than any other factor, which makes water treatment so specialized a technology.

The users

There are two broad bands of water users, domestic and industrial, and it will surprise many people to know that the industrial use of water far outstrips the domestic use. The existence of two main types of user leads to problems with definitions. For instance a power-station chemist responsible for supplying pure water to his boilers has a concept of what is pure very different from that of his local environmental health officer who can be called to check pure water supplied for drinking purposes.

With increasingly high standards of purity being imposed upon the municipal authorities for drinking water and increasingly tight consent parameters being imposed upon both municipal sewage and industrial effluent discharges, the cost of using water is increasing dramatically. The economics of water management both municipally and industrially are changing and the manager who ignores his cost/use options may do his organization an expensive disservice.

The types of water treatment available are applicable in most instances to either municipal or industrial supplies but volumetric requirements are widely different and the municipal industry has tended to be dominated by civil engineers who have to store and distribute over wide areas large volumes of water. Industrial users have tended to use chemists for controlling their water requirements because they operate in a smaller geographical area with lower volumes and tend to need a different kind of purity.

The cycle

All water in a system starts somewhere and ends somewhere. During its voyage it may go through phase changes, become part of another product, be contaminated, cleaned, re-used or discarded. The cycle will normally figure somewhere in the following flow chart.

The two most readily defined waters which are used for potable and industrial purposes are borehole and surface waters. Borehole waters are abstracted from shallow or deep beds and their characteristics derive from their geographical position relative to saline waters, natural aquifers and industrial and agricultural discharges. In some areas there is an unacceptable level of dissolved ferrous and manganese hydroxides which can precipitate on contact with air. In areas of East Anglia and Staffordshire (amongst others) some borehole waters contain unacceptable levels of nitrates. In most areas borehole waters tend to be high in calcium and magnesium hardness. On the whole, however, their treatment for potable purposes requires a lower level of capital expenditure than the surface waters which are subsequently discussed. This economic fact of life has been considerably disturbed by standards imposed within the EC which must lead to considerable expenditure on water supplies to remove impurities which a decade ago were not considered a health risk.

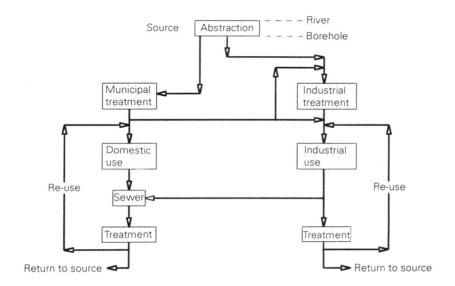

Figure 1.1.   The water cycle

Part 2
PRETREATMENT

2   Pretreatment of surface waters

Surface waters

These waters are abstracted from lakes and rivers and suffer from high organic loadings generally, as briefly explained in Chapter 1. They will also contain a degree of suspended and colloidal matter which may well vary considerably over the seasons. The peatiness of some of these waters gives them a colour which may not be acceptable for domestic use and the organic content makes them unsuitable for many industrial applications.

The task of the municipal authority using this water is to ensure that the water is free of harmful bugs and not offensive in smell, taste or appearance. In recent years much closer definitions have been put on drinking water quality by the EC and closer monitoring of quality is inevitable.

If the responsible authority needs to remove suspended matter and colour it will in so doing take the first step along the treatment path for waters used by the industrialist. The treatment will normally be coagulation/settlement/filtration. For municipal authorities this treatment plus disinfection may be all that is required, but for the industrialist it would normally be followed by further treatments including ion exchange, and therefore pretreatment would be a closer definition for him.

The impurities in water

There is a group of materials all of which must be at low level for public supply;

① ②
dissolved organic matter, organic and inorganic colloids, and organic and inorganic suspended matter. For subsequent ion exchange treatment by the industrialist, the allowed level of these impurities is at least as stringent as for public supply; for other industrial uses, of course, the criteria vary widely.

All the impurities in this group will be found in natural surface waters and their origin may be natural or man-made, or both. It is possible to make arbitrary distinctions between dissolved, colloidal and suspended material but these turn out to be rather arguable definitions of particle size. It is really best to look at these impurities as a continuous series of similar materials.

Organics Natural organic matter derives mostly from decomposing vegetable matter. For our purposes, the most important constituent is a group of large weakly acidic molecules. They give the water a yellow to brown colour, depending on their concentration. Here we shall refer to this class of material as humic acid. The importance of humic acids is that they are taken up on anion exchange resins because of their acidic properties and are then extremely difficult or even impossible to regenerate off again. Porous anion resins have been developed to be more tolerant to humic acids but none are absolutely resistant to this kind of fouling. The porous resins are more expensive than the standard gel type. Some grades are made specifically to be used as adsorber resins upstream of the main ion exchange plant, which raises an even greater capital cost.

There are therefore three practical ways of coping with humic acids: leave them and suffer the consequences, which would be a reduced working life and lower treated water quality from an ion exchange plant; use porous resins either as working resins or as adsorbers; or remove humic acids in a pretreatment plant. Similar remarks would apply to detergents and other man-made organic impurities.

Some classes of organic matter are removed neither by pretreatment of the normal kinds nor by ion exchange. Quite often these are harmless but if they appear in very high pressure boiler feed then the organics break down in the boiler and form (among others) acetic acid and CO_2 and so contribute to corrosion. The total removal of organics is also important in water for semi-conductor manufacture, where the final wash water has to be very pure.

Colloids The biggest humic acids are within the colloidal size range, but usually colloids are clay particles. Surface forces attract them to brand new ion exchange resins where they soon build up an inert coating, after which colloids will pass through ion exchange plant.

Pretreatment in the wider sense

A great many processes can be used to pretreat water. The most important of these are the coagulation processes. Because of the widespread use of coagulation, 'pretreatment' has come to imply that coagulation is involved. Before we turn to pretreatment in this narrower sense, let us briefly deal with some other methods of pretreating water.

- Coarse, heavy suspended matter can be settled out in tanks or taken out in coarse strainers. Hydrocyclones remove sand very efficiently; they consume some power but the plant is cheap and compact.

- Finer suspended matter can be taken out by fine strainer filters such as leaf or cake filters. Unfortunately, the materials found in water tend to form a rather impervious filter cake which blinds the filter quite quickly and can be difficult to wash off. One solution is to use disposable filter cartridges, and another is to install precoat filters. Both these can raise a significant running cost, especially the disposable cartridges. Microstrainers are occasionally used in the public sector: they consist of a rotating drum of very fine wire mesh which is cleaned continuously by high-pressure jets of water.

- Still finer suspended solids and suspended oil can be taken out by deep bed filters.

- Some deep well waters contain dissolved ferrous hydroxide. On contact with the air this starts to precipitate as insoluble ferric hydroxide. To perform this precipitation to remove the iron before it can do any damage requires either aeration followed by filtration, or filtration through a catalytic filter medium which promotes the precipitation of iron inside the filter. The same techniques are used for dissolved manganese.

- Activated carbon is an efficient adsorbent for small organic molecules, especially those causing taste and odour, and is therefore finding increasing use for potable water. Large molecules such as humic acids are scarcely taken out by activated carbon because they are too large to penetrate effectively into the pores of the adsorbent. Fortunately, coagulation processes are efficient at dealing with humic acids but less so in taste and odour control, so these two processes tend to be complementary.

- Living organisms may have to be killed if only to stop them breeding in the pretreatment plant (they might even cause non-reactive silica to be created there!). Chlorination is the most obvious method, but free chlorine attacks ion exchange resins so that the water may subsequently have to be dechlorinated, for example by sulphite dosing or by passing it over a carbon bed. Alternatively one might use ozone as a biocide or (on the small scale) ultra-violet irradiation. Growth of algae etc. in ion exchange beds can be arrested or prevented by slug treatment with any biocide which is non-oxidizing, non-ionic, and non-fouling. Formaldehyde fills this bill.

- There is a variety of filters with organic membranes which have very

13

fine pores. Ultrafiltration membranes have pores small enough to remove most kinds of non-reactive silica with a pressure across the membrane of less than 5 bar. Reverse osmosis membranes actually retain not only large molecules but the bulk of all inorganic salts as well, but the rate of flow across the membrane is small and requires pressures of the order of 20 bar, so that reverse osmosis raises quite large capital and running costs. In semiconductor manufacture the volume to be treated is not large and purity is all-important; this field has proved by far the best reverse osmosis application in industrial water treatment. Followed by ion exchange demineralization it produces water of almost ultimate purity.

Pretreatment by coagulation

We have already shown that pretreatment has come to imply some coagulation process, but coagulation is itself a word which has assumed a narrower significance than it has in the academic sense. Coagulation actually describes a process in which small particles in suspension join together to form larger agglomerates, and we shall start off by describing this particular process.

Particles in water, both suspended and colloidal, carry electrical charges on their surface. Most of the commoner materials assume a negative charge. Constant random movement causes particles to approach or collide with one another. If two such particles carry the same charge, then they will tend to repel one another rather than join together. They will only touch if they are on a true collision course which can overcome the electrical repulsion. When that happens, then van der Waal's force of attraction will tend to keep them joined together to coagulate into a larger particle. The rate at which coagulation takes place clearly depends on the frequency of collisions, so that mixing promotes the effect but violent agitation will break the attraction between particles and inhibit coagulation.

In time, even a stagnant sample of water will coagulate sufficiently to form particles large enough to settle to the bottom leaving the bulk of the water largely clear. This process (called autocoagulation, which sounds like a good name for a traffic jam) improves the water quality in large reservoirs but in the small storage tanks common in industry the process is too slow to be of much use.

An obvious way of promoting coagulation would be to try to neutralize or even reverse the electrical repulsion effect: then particles would be pulled together and even near misses would result in direct hits. There are two ways of doing this: by introducing fresh particles with a positive charge or by adding traces of surface active materials called coagulation aids. Adding fresh particles with a positive charge is best done by flocculation, that is by precipitating iron or aluminium hydroxide in the water (see p.16). For the moment let us consider the use of coagulation aids.

Coagulation aids

These aids are highly diverse with only one thing in common: they are all long-chain molecules, carrying electrically active groups along the length of the chain. For example, among the most effective aids are long-chain polyacrylamides, which carry positive charges set regularly along a long thread-like molecule.

It used to be thought that these aids collected electro-negative suspended particles like washing on a clothes line, with the active groups acting like clothes pegs. Very recent research suggests instead that each long chain molecule collapses on to a single suspended particle where, presumably, it covers a sizeable area as the charged thread settles like coils of rope. This area is then electro-positive and will attract and hold the electro-negative surface of another particle, but this other particle may also have an electro-positive patch on it. And so on.

Whatever the true mechanism, it is an observed fact that the most efficient aids have very long chains; those with shorter chains must be dosed in larger quantities for the same effect. The polyacrylamides, for instance, are purpose made to have very long chains and are effective at doses which would usually be below 1 ppm. Unfortunately they are toxic and may not be used for potable water, so water works have to use safe natural materials like tannins, potato starch or activated silica. This last is a form of polysilicic acid which has to be prepared on the plant as it ages quite rapidly. If the preparation and dosing are not perfectly controlled, it is possible to leave high polymers of silica in the water which act as non-reactive silica. All the materials in the safe group above may have to be dosed at ten times the rate appropriate for polyacrylamides.

Coagulation aids have their disadvantages. The purpose-made synthetics are expensive; on paper this may not matter because of the low dose rate but it makes accidental losses expensive. Their long, thread-like molecules are difficult to disentangle, that is, they are difficult to dissolve, dilute and mix. All coagulation aids will foul ion exchangers. Again, on paper there should be no residual aid remaining in the water which goes forward to ion exchange but in practice some always gets through.

Only trial and error will select the right grade and dose rate from the many aids which are available but the possible side effects of residues remaining in the water are even more difficult to predict. The best thing is to avoid using coagulation aids except for effluent or cooling water treatment.

One good example of their use is in steelworks, where the waste water is grossly contaminated with oil and particles of iron oxide of all sizes. On modern works the really big bits are settled out in scale pits, after which the water goes through coarse media filters (often multi-media). Before filtration the water is dosed with coagulation aid. This promotes coagulation within the bed and helps to hold together those particles which have been filtered out. As a result

15

the accumulated solids can resist higher shear forces without being forced through the filter bed, which permits higher filter flow rates and therefore smaller and cheaper filters. The filtered water is used for rough cooling and washing, in which a little residual coagulation aid is harmless.

Flocculation

Iron and aluminium salts react with water to form insoluble hydroxides which come out as a very fluffy floc whose large surface area carries a positive charge. This active surface not only attaches itself to electro-negative particles in the water but it will also attract and hold the acidic (that is, electro-negative) humic acids. Creating the precipitate is called flocculation and, as we see, it serves the dual purpose of promoting coagulation and of adsorbing humic acids. Flocculation is more effective than coagulation aids alone, partly because it creates a very large electro-positive area in the water and also because it greatly increases the concentration of particles in the water and, therefore, the number of collisions taking place.

The precipitation will only take place within a certain range of pH values. Like all chemical reactions at very low concentration, it requires good mixing and a little time. It is favoured by nuclei on which the precipitate can form but is hindered by high concentrations of organic matter.

Control of pH   Iron and aluminium precipitate best at slightly acidic pH: iron at pH5–6 and aluminium at 6–7. (Another way of creating aluminium floc is from sodium aluminate at pH8 or over, which may be done in conjunction with lime softening. Acid flocculation is much commoner.)

The flocculation reaction itself acidifies the water, as for example when using alum. (Alum contains 18 molecules H_2O as water of crystallization but these play no part in the chemical reaction and have been ignored in our equation.)

$$Al_2(SO_4)_3 + 6\ H_2O \rightarrow 2\ Al(OH)_3 + 3\ H_2SO_4$$

Each equivalent of flocculant therefore releases one equivalent of acid into the water, and increases the strong anion content by one equivalent.

In most cases the raw water arrives with a pH which is higher than the best value for flocculation, so that the lowering of pH due to the flocculation is welcome; with luck the amount of flocculant we need will just bring the pH down to the right level. In other cases we may also have to add either acid or alkali to get the pH right. The actual pH change due to the flocculant is also dependent on the amount of bicarbonate buffering in the raw water. Many English waters contain moderate or high amounts of bicarbonate which will restrict the drop in pH, so these waters favour the use of alum because it works better in the higher pH range. On the other hand, Scottish and Welsh waters often contain little bicarbonate and moreover some of them are fairly acidic to start with. The lower pH range of iron is more attractive in these circumstances. It is a lucky coincidence that humic acid comes out best at low pH, and it is just these waters which present the most serious humic acid problems.

The most important general rule about flocculation is that there is no general rule, at least none to which there are not a number of exceptions which confound the theoretician. The only way to find the right chemical conditions is by systematic trial and error, a laborious process which will be more fully described later (see Chapter 5).

Other factors   The pH and coagulant dose are the most important factors, but there are others. As mentioned above, the chemical reaction will not work properly unless the reagents are properly mixed with the water and it will take an inordinate time to go to completion unless it is promoted by gentle mixing and by the presence of nuclei on which the precipitate can start to form. Organic matter interferes with the precipitation and in very low temperatures the rate of reaction becomes very slow.

A really difficult problem for an industrialist, for example, was represented by a cold Scottish loch in which autocoagulation had removed almost all the suspended matter, leaving a water of high colour (that is humic acids) and low turbidity (suspended matter). Another occurred in Ireland where the raw water from a public supply had been partially treated to remove suspended matter but leaving copious amounts of organics in the water.

3 Sedimentation

We have now dealt with the theory of coagulation and how to get a good floc. In this context, good means easily removed, whether by direct filtration, sedimentation or flotation. If the water contains little floc, and that of a good filterable quality, we might get a reasonable filter run by passing the water straight on to deep bed filters, without removing part of the floc beforehand and this subject will be discussed in Chapters 6 and 7.

Usually, however, most of the floc is removed before sand filtration, and the commonest process used is sedimentation.

The principles of sedimentation

The basic theory is simple enough – floc particles are heavier than water and will settle. As they are both small and light, the speed with which they fall through water quickly reaches the terminal velocity, when the force of gravity pulling them down equals the frictional force in passing through the water. If we assume laminar flow (which is reasonable enough for these small particles and low velocities) then Stokes' Law applies:

$$v = \frac{gd^2(S - 1)}{18.U}$$

Here v is the terminal velocity, g the gravitational constant (omitted if the values are expressed in SI units) and d is the particle diameter. $(S - 1)$ is the particle density minus the density of water, so $(S - 1)$ represents the effective weight of the particle as it falls. U is the kinematic viscosity of water.

The quicker the floc particles come down, the better for sedimentation, so we want to arrange for v to be as high as possible. All the terms in Stokes' Law except g can be improved towards this end.

Improving the terminal velocity

Floc is only slightly denser than water, so that $(S - 1)$ is a small difference between large numbers and a very small change in S will be noticed in $(S - 1)$. Floc density is largely fixed by the water chemistry but in optimum conditions, and with coagulation aids, the particles in the floc can be bound more tightly together and so raise S. Temperature change affects both floc and water together and therefore results in little change of $(S - 1)$.

On the other hand the kinematic viscosity U falls with rising temperature. (Sedimentation plant operators know they tend to have most trouble in winter when the water is cold.) In the Scandinavian winter, raw water may be drawn from under the ice, at 0.5°C, and Scandinavian waters tend to be difficult anyway. For boiler feed water, when the water has to be heated sooner or later, their normal practice is to preheat to 20°C, often using waste heat. The 78 per cent increase in U between 0.5°C and 20°C is worth having.

Clearly the most rewarding variable to attack would be the diameter d, because the terminal velocity v varies as d^2. Hence the preoccupation with particle size growth and the importance of good coagulation to get a 'good' floc. The other variables are of only marginal importance compared with this.

Types of sedimentation plant

The main object in sedimentation plant design is to reduce the time necessary for particles to settle. So far we have dealt with ways of increasing v, the sedimentation velocity. Another approach is to reduce the distance through which the particles have to travel.

This approach has resulted in the development of two basically different groups of designs. The classical sedimentation tank has a large unencumbered volume in which the floc has to settle through distances which may be of the order of a metre. In lamellar sedimentation designs, floc particles only travel distances measured in centimetres.

Lamellar designs

These designs make the water and floc flow through some baffle system which divides the water flow into a large number of parallel streams with a small vertical dimension. There are a number of such devices such as bundles of submerged tubes, or parallel plates, or a honeycomb structure, or corrugated asbestos sheeting.

19

These all have the same essential features: they create parallel water paths which divide the tank into a number of small passages (small, that is, in the vertical dimension; it does not matter too much how wide they are). These are set at an angle in the tank such that the floc settles on to the bottom surface of each passage and forms a layer of sludge which then slides down the passage floor while the water continues to flow up and out through it. At the bottom end of the passage, the floc drops out into some kind of common receiving chamber from which the accumulated sludge can be taken away. Figure 3.1 shows this basic arrangement.

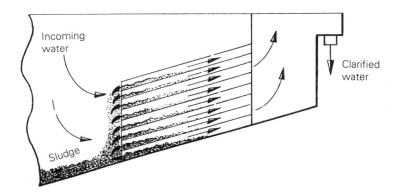

Figure 3.1. Schematic section of portion of lamellar sedimentation tank

The designer's problem is to combine a hydraulically successful arrangement with a structure which is robust, easily maintained and cheap. Lamellar tanks are naturally more expensive than classical tanks of the same outside dimensions but should make up for this by a higher output. This means that a lamellar tank can produce more water than a conventional tank of the same size. But the lamellar tank is not necessarily cheaper per unit throughput. In practice, lamellar designs have been most successful for uprating existing tanks, or in applications where the tank size has to be limited. Where there are no restrictions, the classical designs seem to be holding their own.

Hydrodynamics of sedimentation tanks

So far we have considered the rate of fall of particles in stagnant water, but in an actual tank the water, too, must be on the move. It can either come in at the bottom and go vertically upwards or go across from end to end.

Simple horizontal settling tanks are often used for removing grit or sand, that is in cases where we are not trying to get a floc to grow to its maximum size. On the other hand, vertical flow can be used to promote further coagulation and growth of floc.

In a vertical flow tank, the net settling velocity is the terminal velocity of the particles minus the upflow velocity. (Obviously the upflow must be less than the terminal velocity or the floc would be carried up and out.) Again we have a difference between two low numbers and therefore small improvements in the settling velocity will greatly improve the performance of the tank.

We calculate the rise rate (or upflow velocity) by dividing the flow by the area of the tank. This gives an average value which is not necessarily true at every point in the tank. However actual velocity at any given point must not vary widely from the average, and therefore even flow distribution is very important. A classical cause for uneven flow is excessive temperature rise in the incoming water. What happens then is that the bottom of the tank receives warm water at a lower density than that of the cooler water already in the tank. The resulting convection currents create rapid upflow in part of the tank (and downflow in another part) so that the contents of the tank are turned upside down, bringing a mass of settled sludge to the top. Manufacturers of proprietary tank designs normally prescribe a maximum of 1°C temperature rise per hour.* This aspect calls for very careful temperature control if the incoming water is preheated.

Lamellar designs are less sensitive to temperature change — the residence time of water in the apparatus is shorter, and the baffle system in the tank reduces the onset of convection currents.

The sludge blanket

So far we have considered only how individual particles settle in water. In a horizontal tank, particles do settle as individual bits, each tracing a trajectory like a cannon ball fired horizontally from the top of a cliff (see any A-level applied maths paper). The bigger bits strike nearest the entry, the smaller ones have a longer trajectory. The only interference between particles is the occasional capture of a slow-falling bit by a heavier one falling on top of it. This description applies more or less to what happens inside a lamellar sedimentation device.

A vertical flow tank has the disadvantage that the net settling rate of the floc

* A plant built in the East receives its raw water through a long overland pipeline. The morning sun on this raises the temperature by 20°C between sunrise and noon. When first commissioned the plant failed because at eleven every morning the contents of the tank turned themselves upside down. The problem was solved by mixers in the raw water storage tank which avoided short-circuiting of the storage volume and so slowed the temperature rise.

is its settling rate in stagnant water minus the rising velocity of water at any given point. The importance of vertical flow, however, is that the falling floc meets fresh particles coming up, resulting in fresh collisions, further coagulation and further particle growth. Hence the use of vertical flow in settling iron or alum floc, where further coagulation is beneficial.

The classical upflow design is the hopper-bottom tank. This is an inverted pyramid or cone. Water enters near the bottom and its rise rate decreases as the cross-sectional area of the tank increases. At some level the rise rate equals the settling rate of the floc, and floc builds up into a sludge blanket of very high floc concentration. Fresh particles attempting to pass through the blanket get caught up in it and coagulate with existing structures there (or, in mathematical terms, the high concentration of floc in the blanket leads to a high probability of collision and coagulation). A good sludge blanket permits the use of higher rise rates and yields a more concentrated sludge for disposal.

Improving on the sludge blanket

The sludge blanket principle can be carried further, but detailed ways of doing so depend on the application. The main two groups are industrial and municipal designs. As a very crude generalization, municipal plant is large, conservatively sized and (whether from prejudice or to suit criteria in adjudicating capital expenditure) has a minimum of moving machinery, while industrial plant usually treats a smaller flow and is built with an eye to low capital investment.

Industrial sedimentation tanks mostly improve on the simple hopper bottom design with sludge recirculation, using paddles or impellers to take old sludge and contact it with the incoming water. On the very large scale these paddles could give trouble with excessive tip speeds unless the design uses a multiplicity of units, and altogether this kind of design fits better into the industrial sector. There are a number of proprietary designs on the market which vary in detail rather than in principle.

Many municipal schemes are still being designed with simple sludge blanket tanks but one interesting design is worth singling out. Its manufacturer describes a tendency for sludge blankets to form cracks which present a preferred path for water and floc to slip through. The blanket is consolidated by imposing a pulse on the incoming flow by applying an intermittent vacuum on a chamber in the inlet line. The flow rises to peaks of five times the mean flow which consolidate the blanket. This device (which has no large mechanical moving parts) can sometimes also be used to upgrade an existing tank. Figures 3.2 and 3.3 show the basic arrangements of the two designs.

Plate 1.   Typical clarifier in industry

Plate 2. Raw-water clarification system at major power-station

Sludge removal

Sludge contains very little dry material — some plants cannot do better than 0.5 per cent solid matter in the blowdown and 1–2 per cent can be considered quite good. Sludge disposal thus represents a sensible loss of water and chemicals, and it often creates an effluent problem. Sludge concentration is therefore important. (Sludge is not toxic, but it is an ugly discolourant. If it settles to the bottom of a lake or stream it can smother the natural organisms there, which must be avoided. The alternative to putting sludge into the drain is to lagoon it for further thickening and dumping. Obviously this becomes less expensive if the plant itself yields a sludge high in dry solids.)

A tank creates sludge continuously but the blanket stores it, so that blowdown can be continuous or intermittent. If blowdown is inadequate, sludge accumulates and eventually passes over the top and clogs the filters. If

Plate 3. Clarification plant showing solids recirculation clarifier with filters

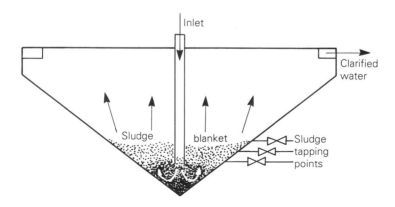

Figure 3.2. Schematic section of hopper-bottom sedimentation tank

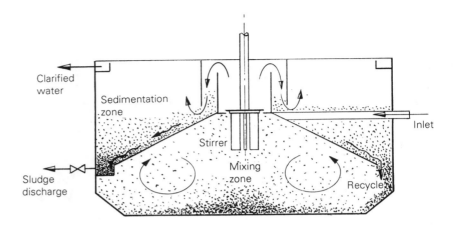

Figure 3.3. Schematic section of sludge recirculation tank

blowdown is excessive the only drawback is that the sludge concentration falls (or, in the limit, the whole blanket is destroyed). Control therefore boils down to blowing down at the lowest practical rate, an uneasy compromise between operator's convenience and the economists' and ecologists' requirements.

One simple method is to have a sludge valve intermittently opened by a variable timer, set by the operator to match the plant's performance. Automatic control of de-sludging, for example by photoelectric cells, has not caught on because of the difficult conditions, in which most devices so far tried have failed. One interesting development (still in the development stage) is to collect the sludge in submerged containers which are emptied when their weight reaches a preset value. If this or any other automatic device is commercially successful it will result in higher sludge concentrations becoming practical.

A heavy sludge can thicken to a stiffish paste and cease to flow freely. Then a sedimentation tank will need scrapers to collect the sludge rather than relying on a sloping bottom to bring the sludge to the blow-off point. The blow-off lines themselves may need automatic flushing to keep them clear. Alum and iron flocs, however, do not usually form very stiff pastes.

4   Flotation

We have described plant for sedimentation, which is the commonest process for removing floc between coagulation and filtration — until recently it was the only one available. Flotation has now become a possible choice for conventional water treatment applications and it can be considered a potential alternative to sedimentation. We still have to see, however, to what extent and in exactly which fields flotation will establish itself. The process has come into conventional water treatment via a number of applications in effluent treatment, and the picture is not made clearer by the fact that there are a number of different flotation processes for which differing and conflicting claims are made.

The basis of flotation

If we attach a gas bubble to a particle of suspended matter we can create a composite particle which is lighter than water and will therefore float to the surface. The exploitation of this phenomenon started in the mining industry decades ago. It turns out that bubbles will attach more easily to some minerals than others, so that flotation is actually used as a selective process to concentrate finely ground ores. In effluent and water treatment, by contrast, it is essential that all the suspended particles should become attached to gas bubbles so that they all float out of the bulk of the liquid.

This means that the process plant must provide enough bubbles for at least one bubble to become attached to each particle. In turn this means that the number of bubbles provided, multiplied by the probability of a bubble becoming attached to a particle, must be at least equal to the number of particles in the water. There are two more potential variables which we can ignore in conventional water treatment, where we are dealing with iron or alum floc. One is the minimum amount of air necessary to float a single particle; these flocs are so light that in practice a single bubble of the smallest size which can be readily produced will give enough buoyancy. The other is the degree to which bubble and particle tend to stay together once they have been brought into contact, and again in the case of these flocs this presents no problem. The important variables are therefore the number of bubbles and the probability of collision.

The cost of introducing bubbles into water is one of the main cost variables in flotation, and for a given process it depends largely on the mass of gas introduced. But the mass as such is not at all important; in this application we are interested in the number of bubbles. It is therefore advantageous to produce the smallest possible bubbles in order to get the largest possible number of them for a given mass of gas.

Small bubbles have the additional advantage that their terminal velocity (controlled by Stokes' Law) is small. This increases the time during which they are available for collision with floc particles, and when contact occurs the approach velocity will be moderate so that the chances of attachment are good. By contrast, very large bubbles are actually harmful: they rise so fast that the violent turbulence which they create can break up floc or knock bubbles off it.

Every argument therefore points to the plant being designed to create the smallest possible bubbles. (In theory it might be possible to create bubbles which would be too small but in practice this limitation can be ignored.) Moreover plant designers try to arrange for any large bubble which might form accidentally to be diverted out of harm's way.

Rise rate in flotation

The movement of the combined particle formed by a bubble attached to the floc particle is also controlled by Stokes' Law. Once again the effective weight of the particle is $(S-1)$ where S is the effective density of the particle and 1 represents the density of water. Now that S is less than 1, the term $(S-1)$ becomes negative, to show the net force is upwards. The benefit of flotation results from the fact that the difference $(S-1)$ is quite substantial and that particles therefore float to the surface much faster than floc sinks in sedimentation, where $(S-1)$ has a very small (positive) numerical value.

Flotation is therefore a much more rapid process than sedimentation, with the result that flotation tanks are much smaller than sedimentation tanks of similar throughput. The total hold-up time in a flotation tank is typically about half an hour, and is a tenth to a quarter the time required in a conventional sedimentation tank.

Coagulation before flotation

This short residence time has important consequences. Above all it means that there cannot be enough time for any significant coagulation or floc growth to take place in the flotation tank itself, whereas in conventional sedimentation tanks designers have plenty of time available, which they utilize in such devices as sludge blankets or sludge recirculation.

Flotation will only succeed, therefore, if the floc is fully formed by the time it enters the flotation tank. This means that a flotation plant needs a separate, carefully designed coagulation section preceding the flotation tank itself. In sedimentation, on the other hand, it is very common to let the tank perform both coagulation and sedimentation in one piece of equipment.

Coagulation plant

We earlier (see p.14) described coagulation and its two main requirements. These are, first, that the chemicals and water must be mixed in such a way that the product (power input × residence time) reaches a certain minimum which is empirical. Second, this mixing must be carried out in such a way that the shear imposed on the water must not break up the floc as it coagulates, which means that it must match the floc strength at various stages of its formation.

Conventional purpose-built coagulation plant therefore consists of a flash tank in which water and chemicals are mixed quickly in high turbulence (which is possible because at the very first stage no floc has formed as yet), after which the water passes through a series of tanks fitted with large slow-moving paddle stirrers, with the stirring rate tapering off progressively as the floc becomes bigger. Ideally each of these tanks has an independently controlled variable-speed stirrer in order to give the greatest degree of control.

Depending on how difficult the water is to coagulate, there will be two, three or four of these stirring stages and the total hold-up of water in the coagulation stage may be up to half an hour. This means that the hold-up can be as great as it is in a flotation tank itself but the overall hold-up of the whole plant is still an hour or less.

The low hold-up compared with conventional sedimentation plant gives rise to the main differences between the two processes. It means that flotation plants respond much more rapidly to changes in conditions whether in the water or in the chemical dosage — this may or may not be an advantage, depending on the application. It also means that flotation can be used more successfully on an intermittent basis — a sedimentation tank will take hours after start-up before its treated water quality is satisfactory, while a flotation plant should be able to produce good water almost at once and will settle down to stable operation in about half an hour.

Experimental work shows that the empirical minimum value for (power input × residence time) holds quite well for alternative methods of mixing. This means that if satisfactory methods can be found of getting more power into water in a shorter time, then the hold-up of the coagulation section could be reduced. One device which works well on the pilot scale, for example, is a bundle of tubes and U–bends, in which the turbulence is created by reversing the direction of flow. It seems that this is one way of getting better mixing without excessive turbulence and so reducing the necessary hold-up. It also benefits from the fact that the water moves along a well-defined path, while in the mixing tanks there is always a lot of back-mixing. Unfortunately this idea is not particularly cost-effective and of course it provides no means of fine control, but it does show the lines on which we may see future progress.

Coagulation chemistry

The chief object in coagulation chemistry, as far as flotation is concerned, is to get the smallest number of floc particles to be floated off, so that we need the smallest number of bubbles and in turn economize on the cost of introducing the gas into the water. As far as achieving the object of treating the water is concerned, of course, we also need a certain minimum coagulant dose in order to trap and remove the suspended matter and the organics.

It now becomes clear why sedimentation is reasonably tolerant of coagulant overdose but flotation is not. More coagulant produces more floc. In sedimentation the extra floc particles settle harmlessly enough; the penalties come only from using a little more coagulant chemical and creating a little more sludge. In flotation, however, the extra floc needs extra bubbles to float it off, which costs money.

Coagulation aids can be useful in flotation but not as useful as in sedimentation. They do a good job in promoting coagulation and reducing the number of particles to float off but their property of increasing the floc density is at best useless and can be positively harmful.

Flotation processes

So far we have discussed flotation in general but there are many ways in which bubbles could be introduced into water, to result in processes with different characteristics:

- compressed air can be dispersed through fine orifices or sinters;
- a stream of water supersaturated with air at high pressure can be released into the bulk of water being treated to release the air as the pressure is taken off;
- electrolysis of the water can be used to raise oxygen and hydrogen bubbles on the electrodes; and
- the whole flow of water can be pressurized, supersaturated with air and the pressure then released.

Dispersed air

This is the simplest, cheapest and least effective method. It cannot produce small bubbles, and it has no application in conventional water treatment.

Dissolved air

At present this seems to be the most promising process for conventional water treatment. It is therefore the process we generally have in mind in the context of this book.

The normal design uses a side stream of clean water, most commonly recirculated clarified water, at a flow which is about 5-10 per cent of the throughput. (At start-up it may be best to use town mains or some external source of clean water.) This side stream is pumped at high pressure through some device in which air is dissolved in the water. Ideally, this device should bring the water near to saturation. If it does not, then it will be necessary either to work at higher pressures (involving higher cost) or to recirculate a larger flow in order to obtain the necessary mass of gas to yield the necessary number of bubbles.

A conventional packed tower for dissolving air is capable of achieving near-saturation. It is supplied with compressed air from a receiver and compressor, which maintains the necessary pressure and automatically makes up the amount of gas taken away by the water. There are other devices which use some kind of direct injection of air, but on the whole these do not achieve the same degree of saturation, though they are cheaper to build.

Float tanks The float tanks using dissolved air do not vary in principle. They are, on the whole, rather simple tanks of about 1.5m depth, with a water path of perhaps 4m. The width of the tank is made to vary according to the throughput rating; alternatively some tanks are circular with a centre inlet and a peripheral outlet.

One ingenious design places the float tank above a conventional deep filter bed so that the clarified water sinks straight through the bed. This means that the flotation tank effectively becomes the filter rising space, and, when the filter is backwashed, the whole flotation plant is washed with it.

Figure 4.1 shows the cross-section of a typical flotation tank. The water inlet and outlet are both near the bottom of the tank; the inlet has to be engineered carefully to give good distribution across the tank without creating turbulence which could break up the floc. Usually there is a sloping baffle near the water inlet to guide the water towards the surface. The recycled and saturated water is fed into the main water stream through a series of nozzles.

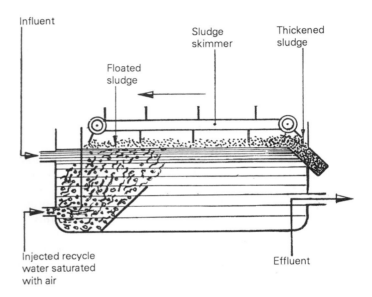

Figure 4.1. Dissolved air flotation tank

The type and number of nozzles vary considerably between different designers' plants. The nozzle must be capable of producing very fine bubbles without transmitting undue turbulence to the bulk of the water, and it must withstand clogging and corrosion/erosion. Some designs use fixed orifices to release the high-pressure recycle; others use a modification of the needle valve.

The positioning of the nozzles must ensure good distribution of the bubbles through the bulk of the water and they must go near the bottom of the tank in order to maximize the possible time for contact with floc particles as the bubbles rise to the surface. If there is any tendency for large bubbles to form, these should be taken away at a place where they cannot interfere with the float which accumulates on the surface of the tank. The tank must not have any overhanging obstructions under water level which could accumulate air-laden floc. In time, such accumulations release their air and fall back into the tank in large lumps which cannot be re-floated and which will therefore pass on with the clarified water.

For good distribution, there is usually a nozzle about every 30 cm of tank width. One convenient way of controlling the flow of recycle is to treat the nozzles as fixed orifices even if they are in the form of needle valves and to vary the water and air pressures supplying the flow.

Electroflotation

Here the bubbles are created on electrodes in the bulk of the water; the plant therefore needs pairs of electrodes with a small, fixed gap between them, through which the DC passes, using the water as electrolytic medium. The ohmic resistance of the water represents a major factor in the power cost and the process therefore becomes very expensive in electrical power unless the water has a fairly high conductivity. This characteristic makes electroflotation more suitable for effluent than for conventional water treatment.

The DC is produced by a transformer rectifier whose cost is roughly proportional to the current required. The electrodes also follow a similar cost pattern. This pattern is different from the bubble-producing gear in dissolved air plant, where the cost of the dissolving plant might vary only as the square root of the mass of air to be dissolved, so that for the very small flows the cost becomes proportionately large. Electroflotation might therefore find application in very small plants for which dissolved air gear becomes disproportionately expensive.

Electrodes are capable of producing very small bubbles in unlimited quantity and without causing any turbulence. This is no particular advantage in conventional water treatment where only a moderate amount of floc has to be removed but it makes electroflotation a good prospect in effluent problems where a very large number of bubbles is needed.

Different designs use all kinds of electrodes. Electrodes are subject to severe corrosion and must either be made as consumable spares, or use expensive materials of construction. Either way the electrodes are a serious cost factor.

Flotation tanks for this process can resemble tanks for dissolved air processes, but there is a large variety of designs.

The static head process

This process uses a shaft about 10m deep, notionally arranged as a U-tube (see Figure 4.2). The water flows in and down one leg and up and out of the other; at a depth of 10m the water is then under about 1 bar static head, but the pumping cost is minimal.

An excess of compressed air (that is, at slightly above 1 bar) is dispersed into

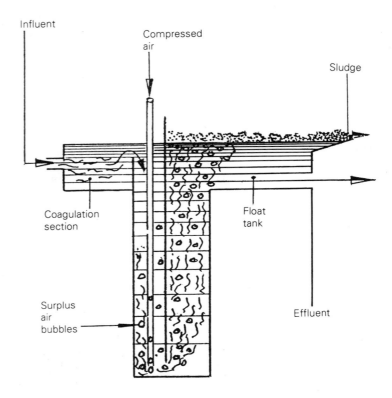

Figure 4.2.   Flotation using static head

the water near the bottom of the down leg; the bubbles rise against the downcoming water and dissolve under the static head. Any surplus air rises to the surface and is lost.

The supersaturated water tends to lose dissolved air as the static head falls as water rises up the upflow leg. Bubbles form on the nuclei provided by suspended particles in the water, and therefore this process effectively avoids the problem of generating fine bubbles and bringing them into contact with the suspended matter. The principle is rather elegant, but practical limitations seem to have restricted this process to treating biological wastes.

Sludge and sludge removal

The sludge forms as a float on the surface of the flotation tank. Compared with sedimentation, the solids content of the float is high, partly because of the greater force pushing the rising particles into the accumulated mass of sludge and partly because air bubbles in part replace the interstitial water.

In the float there is a tendency for the small air bubbles to coalesce in time and leave portions of sludge without attached bubbles. This does not matter as long as the float is not disturbed because the fresh-floating particles continue to come up from underneath and hold up the whole layer of float. While the sludge is being removed, however, there is also a tendency for bits of sludge to detach themselves, sink back into the tank and pass out with the clarified water. Sludge removal must therefore be done with care.

Most designs have continuous scraper gear skimming sludge off the surface and on to an inclined beach, from which it drops into a hopper or trough. The skimming rate is adjusted so that the float builds up to a thickness of 2–3 cm; on the one hand this permits the sludge to be taken away at a reasonable solid concentration (at least 1 per cent) but on the other it means the float has not been left long enough for a serious amount of material to drop out when it is disturbed. Another method is to allow the sludge to build up and periodically to raise the water level in the tank so that it overflows a weir, carrying the sludge with it.

In some conventional water treatment operations the coagulant dose is quite small (especially when compared with the amounts used in effluent applications). The sludge then accumulates quite slowly. Where this is the case, it is possible to let the float accumulate for several hours at a time and remove the sludge on an intermittent basis. If the float can be left to accumulate for more than eight hours, say, it might even be possible to control desludging manually and save the expense of automatic gear. The float becomes extremely stiff and may be up to 10 per cent dry so that it will not flow easily and may need helping along, especially near the walls of the tank (to which the float tends to adhere). This can be done with water sprays, or manually. This kind of desludging leads to a slight loss of quality in the clarified water for ten or twenty minutes, but at such infrequent intervals that the additional load on the following filters is effectively very small.

Flotation vs sedimentation

We might now compare the dissolved air process (which is the main contender for conventional water treatment applications) with the conventional sedimentation tank which is at present the standard apparatus for floc removal.

The main difference between the two processes is the short hold-up time in flotation. Flotation therefore uses much smaller tanks and its capital cost overall ought on that account to be lower. It takes up much less space, and for small sizes can be built skid-mounted or even transportable. On the other hand, the float must not be exposed to rain or wind, both of which might break it up, so that the plant must be housed in a light shelter at least.

The short hold-up makes flotation plants much more responsive to changes in conditions, as already mentioned. On a water liable to rapid changes, such as a small Scottish stream, this could mean almost continuous manual attention to the plant when a period of rain follows one of drought. A sedimentation plant in the same circumstances is also likely to give trouble but it will not require such continuous attention for the simple reason that its response time to changes is so slow that there would be very little point in sitting on it continuously. On the other hand, flotation plants can be started and stopped almost without delay, which might be very important in suitable circumstances. While flotation plants may show a lower capital cost for performing a given task, they inevitably show a higher operating cost. This is due to the cost of high pressure pumping of the recycle and air. There may be some small saving due to the fact that flotation plants may be thriftier in coagulation chemicals than sedimentation plant, but this will not cancel the increased cost due to pumping.

Flotation can produce as good a quality of clarified water as sedimentation, or sometimes even better. 'Can' means that the comparison is made between the products of two well-operated plants. It does not take into account the fact that in many cases, sedimentation plants can tolerate a greater degree of the ignorance, neglect and maloperation which is unfortunately the fate of many water-treatment plants. When well operated, flotation is capable of giving a better product when treating the difficult thin, cold, coloured waters of Scotland and such countries. This is why the use of flotation for water treatment received much of its impetus in Scandinavia where these waters are common. On the other hand, waters of high turbidity are more suited to sedimentation. Algae removal (which is difficult by any other means) is performed rather satisfactorily by flotation.

Flotation produces a smaller bulk of sludge, and with care is capable of producing a very dry sludge. In some cases this can be a very important benefit.

Summing up: flotation is a relatively new process, especially in the field of conventional alum and iron floc removal when treating natural waters. It is still not quite clear to what extent it will find an economical application in the field in which sedimentation has served at least adequately for so long. Algae removal and the treatment of thin, coloured waters by flotation both look promising. It seems as though the dissolved air process is the most suitable but, as the process is still being developed, something new may yet emerge.

5   The jar test

We have described the surface electrical charges carried by all suspended particles and how these can be modified in order to persuade particles to coagulate together and form larger flocs. In water treatment, 'coagulation' is generally used in a narrower sense. It describes a process in which iron or aluminium hydroxides are precipitated in the water so that this flocculant precipitate can coagulate together with the suspended and colloidal matter which was previously in the water and can also adsorb organic matter on its surface. We now come to discuss coagulation in this narrower sense.

Testing for coagulation

While the basic principles of coagulation are well understood, its practice is full of uncertainties and exceptions. The main reason for this is that the impurities in water vary in a great many aspects which no simple measurements can characterize. The variables are size distribution, shape, density, surface charge of suspended and colloidal particles, and the concentration and chemical nature of dissolved organics. These intangibles interact with one another and therefore those measurements which we can make will only be a vague guide to help us use our experience of similar waters or impurities of a similar origin. To predict with confidence how a coagulation process will work we have to simulate it experimentally.

In such experiments, we shall have to remember that the properties of a sample of water change with time. We have described how suspended particles will coagulate in time, a process which is accelerated by gentle agitation. An old sample, or one which has spent a few hours in the back of a moving car, will be partly autocoagulated (no pun intended). Tests on it will give only a rough guide to the behaviour of the water as it actually comes.

The upshot of it all is that one cannot design an efficient coagulation plant from a water analysis alone. Tests on aged water samples are better than nothing, but for reliable results tests should be carried out on the water exactly as it will flow into the plant, which means testing on site. Quite often the supply of water which is to be treated is not yet available and testing is impossible. Then the designer has to allow plenty of safety margin in case the water turns out to be more difficult to treat than expected. Efficient design requires testing on site.

The jar test

The generally accepted test method is the jar test, which is, unfortunately, laborious. Research work into easier methods is proceeding but for the present we are stuck with the jar test. It consists of putting samples of water into beakers, dosing, stirring and allowing them to settle, all under standard conditions. This simulates crudely what a plant would actually do, and the usefulness of the tests depends on their being repeated systematically and with care, as a series of planned experiments.

The basic apparatus is simple. It consists typically of a stand of frosted glass lit from below. Six standard 600ml beakers stand on the frosted glass side by side and in each of them there is an identical flat stainless steel stirrer paddle. These stirrers run off a common drive actuated by a variable speed motor. The paddles are quite big, so that good stirring is obtained without very high speeds (a typical paddle has an area of about one-fifth of the vertical cross-section of 500ml water in the beaker). The object of this arrangement is to be able to subject six samples to identical stirring conditions and then observe clearly what happens in the beakers.

Greater sophistication of the kit can improve the consistency of trials. Some kits have a set of stainless steel scoops, one above each beaker and mounted on a common shaft. The chemical dose is measured into each scoop before the trial starts and, by turning the common shaft, each beaker is dosed at the same instant. This avoids the time delay of dosing each beaker in turn. There can be one, two or three such sets of scoops, each set for one chemical — this improvement is particularly useful when investigating the use of coagulation aids because of the critical time interval between dosing the coagulant and the coagulation aid (see p.43).

For carrying out a batch of tests, each beaker receives a 500ml sample and the paddles are started. The chemicals are dosed in, and the six samples then get a 'fast stir' followed by a 'slow stir'. The stirrers are then stopped and the samples allowed to settle.

Normally the six samples in a batch are kept identical except for one common variable, which might be the amount of coagulant, the amount of pH correction, or the amount of coagulation aid. This means that the effect of this common variable can be studied in constant conditions without the interference of slight changes of conditions which can occur between batches of trials. These differences between batches might, for example, be due to ageing, or heating up of the sample standing in a warm laboratory.

Stirring speeds and periods are usually kept standard for a whole series of trials. A typical set of conditions might be: fast stir, 60–80 rpm for 1–4 minutes; slow stir, 5-20 rpm for 15–20 minutes. Each batch of six trials therefore takes half an hour, more or less.

Interpreting the test

After the slow stir the samples are left to stand and floc can settle out. Now the operator has to observe the floc and the supernatant water, helped by the illumination through the frosted glass stand. Ideally there will be a large dense floc settling quickly from a clear and colourless liquid. At worst, the water will end up more turbid than it was before with only a few bits and pieces falling to the bottom.

For each batch of trials, the following should be recorded for each sample:

- *Size of floc.* A standard set of descriptions is used: 'pin-point', extremely fine but visible; 'fine', 0.5mm average; 'medium', 1mm average; 'large', 2mm average.
- *Rate of settling.* Large flocs settle fastest, unless different chemical conditions result in different floc densities. By and large if the floc settles within four minutes of standing, then a sedimentation tank should work fairly well. If settling takes four to seven minutes, a plant will need extra care in design and operation. A floc which takes longer to settle is not really suitable for treatment in conventional plant.
- *Appearance of supernatant liquid.* The last two observations show how easily the floc can be removed after coagulation but observation of the supernatant water shows whether the process has actually achieved what it set out to do, that is the removal of colour and/or turbidity. It also gives an indication whether it is going to be difficult to filter out the residual floc after sedimentation or flotation. The operator records whether the supernatant is clear and colourless, and describes the amount and appearance of residual floc particles.

This last observation can become confused by the release of gas bubbles from the water while the test is being carried out, either because the water was supersaturated with air in a pressure main or because its rising temperature in a warm room brings air out of solution. (Lowering the pH during the jar tests releases CO_2 into the water, but this is in itself not enough to cause bubbles to form.) These bubbles attach themselves to floc particles and stop them settling, or even float them to the surface.

With each batch taking half an hour or more, a series of jar tests is a lengthy procedure. It is therefore particularly important to make consistent observations and to record them methodically. Only then will it be possible to see the true trends when the results are collated.

Jar testing for flotation

In most cases the intention is to remove floc from the water by sedimentation, but it is also possible to float it out with air bubbles, using the sort of effect described above. Jar test kits for flotation therefore have to simulate the process described in Chapter 4. Flotation jar test kits are much more complicated than those for sedimentation, and they are not freely available. However, the standard jar test can be quite useful in predicting flotation performance.

The basic feature of flotation processes is that enough bubbles must be introduced into the water to ensure that each floc particle will capture at least one bubble and so float out. Producing large numbers of bubbles is expensive, and therefore the process works best when the number of floc particles is reduced to a minimum. In other words, for a given coagulant dose level, we want the largest floc particles in order to have the smallest number of them. In most cases this gives the curious result that the best settling floc can also be the best floating one.

The standard jar test will tell us nothing about flotation problems such as bubble attachment but it is an excellent guide to the best chemistry for flotation.

The chemical variables in coagulation

The most important thing in any coagulation process is to get the chemistry right. The variables for this are: the choice between iron or aluminium salts, the pH and the coagulant dose. These three variables are closely inter-related.

Chapter 2 mentioned that iron salts precipitate best around pH 5–6 while aluminium is best at pH 6-7. Alum is the cheaper of the two and less will be needed to get the pH down for its best range. On the other hand, humic acids are removed more efficiently at low pH. A very general rule is therefore that alum is used unless the main problem is the removal of organics.

It also explained how 1 ppm as $CaCO_3$ of coagulant introduces 1 ppm as $CaCO_3$ of acidity into the water. The pH of raw water is almost always higher than the coagulation pH. With luck it is often possible to hit on a coagulant dose rate which gives the right amount of floc at the right pH; without luck additional acid will be needed to bring the pH down further, or an alkali dose to bring it back up again.

Some examples will show how this works in practice. (All chemical quantities are expressed as $CaCO_3$.)

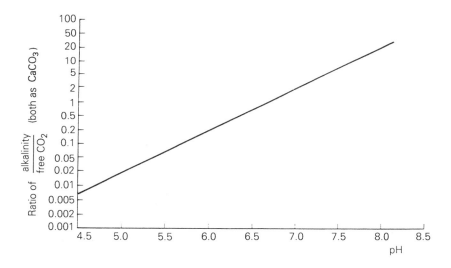

Figure 5.1. Relationship of pH to the alkalinity/free CO_2 ratio

Example 1 Suppose we have a water of pH7.7 with 220ppm alkalinity. Figure 5.1. shows that at pH7.7 the alkalinity:free CO_2 ratio is 10:1, so that this water must contain 22ppm of free CO_2. Suppose we now dose 30ppm of alum. The acidity which this introduces converts 30ppm of alkalinity into 60ppm of free CO_2. Now we have a water which still has 190ppm of alkalinity and whose free CO_2 has been increased to 82ppm. The alkalinity:free CO_2 ratio in it is therefore 2.3:1, which gives us a pH of just over 7. If the water is an easy one to treat, this might give us a reasonable floc, but if it is at all difficult we have to bring the pH down for a workable process. This can be done by increasing the alum dose or adding sulphuric acid. Acid is cheaper than alum but it involves a second chemical with more capital cost and potential trouble. For sedimentation, a higher dose of alum might be beneficial or at least harmless, so this would be the normal solution to the problem. Flotation is not so tolerant of an overdose of coagulant because it increases the resulting number of floc particles (see p.40). For flotation, therefore, we may well have to add acid to the alum dose. If the proportions of alum and acid are likely to be reasonably constant, the two chemicals can be mixed and dosed with a common dosing pump, which is cheaper than dosing them separately.

To sum up this example: the optimum alum dose may just bring the pH low enough or we may need to acidify further, either with more alum or with an acid dose.

Example 2 Now suppose we are treating a moorland water with 40ppm alkalinity and pH7.1. It is a typical moorland water in being highly coloured, and we expect that in order to take out its high content of humic acids we shall have to operate at low pH. We therefore set out with iron (Fe^{+++}) salts – ferric chloride, ferric sulphate, or chlorinated copperas. Suppose we dose one of these at 30ppm (as far as the pH is concerned it doesn't matter which). Following the same logic as Example 1, at pH7.1 we find that the raw water contains 13ppm free CO_2 and after dosing 30ppm Fe^{+++} we shall be left with 10ppm alkalinity and 73ppm free CO_2, which gives us a pH of about 5.8. This ought to be about right for a good iron floc and it ought also to give good removal of the colour.

Example 3 Let us take the same moorland water as in Example 2, but assume that it needs a dose of 50ppm Fe^{+++}. There is only 40ppm alkalinity in the water, so this dose completely removes the alkalinity and the pH drops below 4. In order to get good precipitation we shall have to dose alkali in order to bring the pH up again. NaOH is the most convenient chemical for bringing the pH up again; lime, though cheaper, has to be handled and measured as a solid or a slurry, which is troublesome. However, lime gives good coagulation by itself and often enhances the effect of iron or alum floc. It can be used very satisfactorily together with alum at pH above 7, but then the absorption of organics is poor.

It should be noted here that whatever flocculant is used additional dissolved solids are being added to the water. Where this water has to be further treated for industrial use by demineralization (see Parts 5 and 6) the further treatment plant has to be sized to remove also the additional solids. In cases where very thin waters are being pretreated by the addition of alum or ferric salts the effect on post-treatment plant can be severe.

Coagulation aids

Here, for a change, we have a description which is accurate. Once the coagulation chemistry is right, then the use of aids can be very helpful. On the other hand we have pointed out that they should be avoided if possible in all cases where the pretreated water is going on to a demineralization plant.

Further jar testing, after the chemistry has been established, will show which aid should be used, at what dose rate and at what point in time it should be added. Sometimes the use of aids makes it possible to reduce the coagulant dose and therefore the whole test programme may be very lengthy.

The time at which aids are added is an important variable. We earlier described how aids work by settling on the surface of particles and changing their electrical charge at that point. At the instant that the coagulant begins to precipitate, the vast number of minute particles which appear in the water present an enormous surface area, over which the aid would be dissipated without doing very much good. The aid has a more powerful effect if it is allowed to settle on the precipitate after it has started to coagulate; then the dose per unit area of particles is much greater.

Thirty seconds' delay can make a tremendous difference but even this short delay can be expensive to build into a plant. A highly efficient method of obtaining it becomes available if the raw water comes down a long pipeline. Then the chemicals can be dosed upstream in the pipeline and the aid at the entry point into the plant.

The thread-like molecular form of coagulation aids makes them dissolve and mix with difficulty. Special care must be taken to overcome this property.

Hydrodynamics

The chemistry can only function properly if the chemicals are dispersed uniformly in the water, and the growth of floc after that depends on collisions between particles. Both parts of the process therefore depend on thorough agitation of the water, though it must never be so violent as to tear apart floc particles which have already coagulated.

The violence of agitation, that is, the liquid shear, which floc can withstand depends above all on the floc size. It also depends on the strength of the bond between particles, and in this respect coagulation aids are very effective in strengthening floc. In all cases, therefore, there is no limitation on liquid shear at the instant when coagulation chemicals are added, but after the first few moments the allowable shear falls rapidly. For optimum coagulation the dispersion of the chemicals takes place in a violently agitated flash tank, after which the water passes through successive zones or tanks with decreasingly violent mixing; this is called tapered stirring.

Empirical formulae have been evolved to calculate the minimum energy input necessary for coagulation. They all boil down to the same substance, which is that the power input times the residence time should exceed

43

some constant. This faces the plant designer with an uncomfortable choice: if he economizes on the size of tanks, he reduces the residence time and is obliged to arrange for a high-energy input into a small mass of water. This is difficult without incurring high shear, at any rate locally as, for example, at the tip of a stirrer paddle.

The problem becomes more acute with very large throughputs, because a large diameter paddle at low revs yielding a low average Reynold's number may still give excessive shear at the tips. In this respect the jar test bears little relation to full-scale plant. The tip speed of the paddle in a 600ml beaker is unlikely to exceed 200mm/sec; the same tip speed is reached by a 2m diameter paddle at 2rpm! This is why the rate and time of stirring in the jar test is not a useful experimental variable; it would produce little helpful information about the hydrodynamics needed on the full scale.

Summary

It will be clear from all this that jar testing is a tedious but skilled affair. Starting from scratch, a complete programme of jar testing may take days to complete and, if the water quality is variable (which is the rule rather than the exception with naturally contaminated watercourses), it may be necessary to carry out trials through the full cycle of the seasons. The whole business is empirical and ought to be done by an experienced operator at the source of the water. Still more experience is needed to translate jar test results into probable plant performance. It is all very unsatisfactory but nothing better has as yet been developed.

Here is a summary of general experience in how to get a good floc, though exceptions to many of these rules can be striking:

- Alum works best at pH6–7, ferric salts at pH5–6, but there are many satisfactory plants operating outside these limits.
- For each water and coagulant there is a pH range in which the best results are obtained. In a difficult water this range may be as narrow as ± 0.1pH units.
- For each water there is a minimum satisfactory dose of coagulant. For sedimentation this can be exceeded by quite a wide margin before any adverse effects are seen, but flotation is more sensitive.
- The coagulant dose may by itself give the correct pH, or acid or alkali may have to be dosed as well. Lime is positively beneficial in most cases but NaOH is commoner because it is more convenient.
- Colour, that is high humic acid etc., is best removed at low pH. High humic acid concentrations interfere with the growth of floc and make the water difficult.

- Floc forms best on existing particles. If the raw water is deficient in these, floc should be recirculated or some seeding material such as Bentonite can be added.
- The ideal regime for mixing consists of a flash tank with violent stirring, followed by zones of decreasing shear to promote the growth of floc.
- Very cold waters benefit from preheating.
- Air bubbles should be avoided.
- Constant, skilled supervision is needed to keep a coagulation plant working consistently on a varying water.
- For greatest effect, coagulation aids should be dosed a short time after the chemical reaction has started.
- Coagulation aids need special care in dissolving and mixing.

Even if all the above precautions are observed, there are some waters which may never give a good floc. The really difficult problems arise from violently varying waters such as small watercourses in the Scottish Highlands, or from lake waters which combine high organic content, low temperature, low alkalinity and zero suspended matter. A 'second bite' at waters which have been partially treated is also troublesome. Once again we must stress that this whole subject is full of exceptions and oddities.

Part 3
FILTRATION OF SOLIDS

6   The background to design

Filtration means taking suspended solids out of fluids, a huge field which includes gases and liquids, sometimes with the object of cleaning the fluid and sometimes to recover a valuable solid. Here we restrict ourselves to water treatment, dealing with general principles and practical applications.

The suspended solids in water

Water may carry particles of all kinds, shapes and sizes but we normally use the term 'suspended matter' to cover quite a narrow range. Very large solids such as salmon or pieces of driftwood, which are easily removed on coarse screens, are not usually included in the term. Sand and other particles large and heavy enough to settle out quite easily in a sand trap or sedimentation tank are not normally included either.

A water which has been screened and settled may still contain suspended matter and also colloids. Colloids are particles so small that they cannot be seen under a simple microscope. The agitation due to the movement of the water molecules themselves is enough to keep them permanently in suspension. The usual meaning of suspended matter therefore covers small particles which do not settle very readily but which are bigger than colloids. Filters in water treatment can be expected to take out all or most of the suspended matter; sometimes they are designed to remove colloids as well.

Basic filtration

In its crudest form, filtration is like tea-straining. The fluid is passed through a screen with small apertures. Particles which are bigger than these apertures are held back on the screen. Screen means any solid barrier with small holes in it, and screens can be created in many ways, some of which will be described below.

This simple picture leads to two problems:

- The suspended matter in water is very small and water is relatively viscous (compared with, say, air). This means that very small holes must be used, and any reasonable rate of flow through them will raise a high pressure loss.
- The particles which are held on the screen accumulate on the holes, tending to blind them and raise an even larger pressure loss. Worse still, as the suspended matter accumulates, a filter cake of very fine particles builds up, through which the water now has to pass to get to the screen at all. The hydraulic resistance of this cake is, in normal water problems, much higher than that of the clean screen.

Filterability

If a filter cake is allowed to form, it effectively forms a new screen in which the holes are much smaller than those of the original screen. The cake then dictates the size of particle which passes through, and a much finer filtration results. This phenomenon is called autofiltration. The small pore size of the cake tends to increase the pressure loss, but if the cake remains reasonably open, with a large pore volume, the effect is not too serious.

On the other hand, the suspended matter may be composed of particles which tend to pack closely into one another by virtue of their shape, or because they slide easily over one another, or because they are pliable and change shape against one another. The degree to which they do these things depends on the pressure loss across the cake, which forces them against one another, tending to compress the cake. When a cake compresses, the pore size is reduced, which increases the pressure loss, which increases the force exerted, which increases the compression, which increases the reduction in pore size, which . . . etc. A really bad and highly compressible material is capable of forming an almost impermeable membrane across the screen through which a small trickle of extremely clear water flows.

The filtration properties of the suspended matter which occur in practice cannot be predicted from the basic properties of the material particles: this can only be done with carefully selected laboratory materials. There are however

Plate 4. Chemical dosing and regeneration station at major power plant

various empirical tests for filterability in which the water is filtered on standard screens. It goes without saying (Murphy's Law being universal) that most of the suspended matter which we wish to filter out of water has poor filterability: natural waters contain clay, decaying organic matter, iron and manganese oxides; treated sewage contains slimy organic residues; condensates contain colloidal crud; steel mill wastes contain grease-laden oxides; paper-mill effluent contains clay and cellulose. All these materials are more or less horrible to deal with. The usual way out of this problem is not to use simple screens at all but deep-bed filtration, which is somewhat different.

51

Filter-washing

Compressible materials which squeeze and cling to one another are also likely to cling to the screen itself, the more so as they will have been pressed firmly into the screen by the high pressure loss which they cause. Such materials are therefore not only troublesome to filter, they make it difficult to clean the filter for the next cycle. In fact filter cleaning can often be the most serious problem when filtering materials of this character.

Types of filters

In designing a filter for a specific task, the designer has only limited freedom. He proposes to remove particles by passing the water through orifices of a slightly smaller size and the hydraulic resistance of these holes and of the cake which builds up on them is a property of the suspended material and the water. He has no control over these. His job is to select a suitable type of screen and to design a filter with the largest economical screen area, so that the specific flow rate through the orifices is small and the pressure loss low. A low specific flow rate also means that the build-up of cake will be slow so that cleaning the filter need not be too frequent. On the other hand, if cleaning is easy and efficient then it is possible to clean relatively frequently and keep the build-up of cake to a minimum.

There are two kinds of simple screen filters:

- True screens, such as meshes, sieves, sinters, felts, fibre-wound candles or cloths, held in some device which forces water to flow through. There must be some form of cleaning.
- Disposable screens, which are created afresh for every filter cycle. This class includes paper filters and cartridge filters, which are used chiefly where there is little suspended matter to remove, otherwise filter replacement becomes too costly. More important in this class are precoat filters.

Microstrainers

The only simple screen which is at all widely used in water treatment is the microstrainer. Basically this is a cylindrical drum covered with a very fine wire mesh. Raw water flows into the middle of the drum and out through the mesh by gravity. The drum rotates continuously and brings the mesh under a high-pressure jet which squirts water on to the outside and dislodges the accumulated matter inside the drum. The smallest practicable mesh is 25 microns, so the finest particles cannot be removed by microstrainers, but they are used for removing algae or for rough filtration of treated sewage.

Precoat filters

Precoat filters use a carrier screen with relatively large orifices. The disposable screen is built up on this by depositing a layer of precoat medium on it.

Filter precoat media are relatively coarse materials with a good filterability, such as diatomite or certain grades of cellulose fibre. They are composed of particles of reasonably uniform size so that they do not pack into one another, and they form filter cakes with a high pore volume. Being rigid, they do not compress significantly under flow.

Before the filter run begins, a suspension of filter medium is prepared and is pumped into the filter. The screen is so coarse that some of the precoat material passes through but most of it is retained on the orifices, across which the rigid particles tend to form bridges. The water coming out of filter is milky with residues of medium and is recirculated back to the inlet.

As soon as any part of the filter is covered with precoat the pressure loss across that part is higher than that of the bare filter, so that flow is diverted on to the remaining bare portions. This automatically distributes the precoat evenly across the filter area. The precoat suspension is recirculated until the filter runnings are clear, which indicates that all, or almost all, the filter area is covered with precoat. The thin layer of precoat is now ready to act as the actual filtering screen.

The filter run now starts, and a second layer of filter cake builds up on top of the precoat, this time made up of the suspended solids from the water. There is of course some penetration of suspended matter into the precoat, so that filtration is no longer entirely by simple straining but begins to approach the principle of the deep-bed filter. If the suspended matter in the water has a very bad filterability, it may be useful to go on dosing a small amount of precoat medium into the water upstream of the filter. This additional material deposits itself together with the filtered suspended matter and acts as a skeleton which gives some stiffness to a cake which might otherwise be very compressible. It also provides discontinuities along which water can flow more easily. This practice is called body feed or body dosing.

As the cake builds up, the pressure loss rises. Precoat filters are usually built for quite high pressure losses limited by the cost of the pressure vessels and of pumping rather than by the process itself. Final pressure losses of two bar or more are normal. At a predetermined limit the run is brought to an end, and the two-layer cake washed off and thrown away. The precoat medium is therefore a consumable item which raises an operating cost.

In backwashing a precoat filter we experience the opposite of the automatic distribution which helps achieve an even precoat: as soon as any part of the filter is clean, the backwash water flows more easily through that part than the unwashed portions. Unless special care is taken, there is tendency for old precoat to stay in place and blank off a part of the filter area. Filter washing is usually assisted by high-pressure water jets or air blowing which (hopefully) gets the filter clean. On the other hand, once a filter run has started flow must continue or there is a danger that some precoat may fall off the support. If the flow is intermittent, an automatic recycle pump may have to be installed.

As mentioned above, precoat filtration raises a running cost. To get the most out of the precoat medium, high pressures are used which mean that the filter has to be in a pressure vessel. The design detail varies but must deal with the problem of getting the greatest possible filter area into the pressure vessel. Some filters use a series of filter leaves, and others use candles hanging from or standing on a tube plate to meet this requirement. The space between filter surfaces must allow the precoat and the suspended matter to build up on it and still permit a passage for the water flowing to the surface of the cake. This requirement limits the filter area which it is possible to cram into a given volume.

Filtration in depth

The deep-bed filter is a vessel which holds a bed of a fine granular medium such as sand, through which the water is made to flow. The suspended matter gets caught in the discontinuities which occur within such a bed. When the medium has accumulated as much solid as is practicable, the bed is agitated to make it release the solids which are then washed away in an upward stream of wash water, after which the filter can be put back on service.

The design and operation of such filters is rather simple and their use goes back centuries. Just exactly what goes on inside them, however, is still (in part) a subject for scientific investigation. The description which now follows is highly simplified.

Water flowing through a bed of granules can pass through a very large number of channels in parallel, but it does not have to go through any one of them. Though the main flow is in a vertical direction, the water can flow across the main direction. The channels interlink, open out and narrow down at random. A sand filter therefore represents an infinite number of mini-events of far greater quantity and variety than those in the simpler types of filter.

In such a variety of mini-events, suspended matter can be retained by different mechanisms. The most obvious one is that a particle arrives at a narrow orifice between sand grains in which it gets stuck. This would be similar to ordinary straining and once again it does not necessarily close off the orifice, so that a mini-cake can form on the original particle. At other places, particles will form bridges across orifices, as a precoat medium does, and create a new, smaller series of orifices across which filtering takes place.

Such events are not however the major contributor to filtration. The important characteristic of the sand bed is that it is full of points of very low liquid turbulence and shear, at which particles can easily settle out on to the surface of the sand. Where these conditions occur, particles are likely to settle on top of one another, so that this type of filtering also leads to local accumulation of suspended matter. Once the particles have been brought into contact by settling in the same place, electrostatic forces tend to hold them together.

As the medium accumulates filtered dirt, the free area available for flow becomes reduced and the pressure loss across the medium rises as local flow velocities increase. This leads to increasing shear forces on the accumulations of suspended matter. The rise in shear force may collapse some of the bridges which hold mini-cakes, and it may cause avalanching in the layers of settled particles clinging loosely to one another and to the surface of the sand. When this happens, rafts of coagulated material begin to pass along with the water: at first they will be retained further along in the medium but in the end they may come out of the medium and cause filter breakthrough.

This is a highly simplified view of what goes on inside the medium. There are still some unknown factors such as, for example, the part played by the nature of the surface of the medium and that of the suspended particles. However, this very rough description of the process serves mainly to assist in understanding the ways in which deep-bed filters differ from all other types of filter:

- They are capable of filtering out very fine solids, down to about $2–3\mu$ in diameter.
- Fine sand gives a finer filtration because the size of the channels is decreased, and because far more area of medium is available. On the other hand the flow rate also has a considerable influence on the fineness of filtration — low flow rates cause less shear and allow more and smaller particles to settle in the filter.
- Because particles tend to accumulate at the same place, sand filters promote coagulation. This tendency can be put to good use.
- Deepening the bed increases the available area of filter medium and improves the quality of filtrate, all other things being equal.
- At the end of a filter cycle, the rising pressure loss can force accumulated dirt right through the filter.

Backwashing the bed

Sand filters are always backwashed upwards in order to allow the bed to expand

and so release the accumulated solids. Some other form of agitation is usually used to help release them.

The theory of backwashing (see p.163) shows that the maximum liquid velocity at any point is the terminal velocity of the individual grains. This is quite a moderate velocity and it does not exert very much shear force on the grains: normally the force is not enough to rip the accumulated material from it. Extra agitation is necessary for efficient washing and this may be supplied by mechanical rakes or water jets or by bubbling air through the bed.

In theory, also, a bed on backwash should classify, so that the smallest particles end up at the top and the largest at the bottom (this assumes that their density is uniform). Sand filters are not usually fluidized for complete classification as this takes very high flows. Bit by bit, however, the bed does tend to classify. With a bed of uniform density the finest particles end up at the top.

Upflow and downflow filters

The usual sand filter operates in downflow. Raw water therefore meets the finest sand first and goes out through the coarsest. Most of the filter's work is thus done at and near the top, with the bottom of the bed acting largely as a support. Obviously it would be better for the water to go through the large channels first, and this unsatisfactory state of affairs has bothered filter designers for a long time,

Upflow filtration is an obvious remedy. The first recorded upflow filter was described in 1685, the first year of James II's reign (it wasn't an unqualified success). We will later describe the designs of upflow filter in use today, and their limitations (see p.65).

Multi-media filters

Another way of beating the problem is to use two or more media of different densities so that light but coarse granules are made to classify above a layer of fine but heavy ones. Filter media available include the following:

Material	Typical specific gravity
Polystyrene	1.04
Anthracite	1.40
Silica sand	2.65
Garnet	3.83

It is this variety of materials which means one ought really to speak of filter media rather than just sand. Multi-media filters (which are becoming much commoner) are also described later.

Multi-media filters have one major theoretical problem. If the gradings of the different media are not carefully controlled, then the terminal velocities of the largest anthracite particles (say) will be the same as or greater than those of the finest sand particles. With anthracite and sand this happens when the ratio of particle diameters is bigger than 5:1, and if this ratio is exceeded the two materials will classify together to form a mixed layer of very different sizes instead of separating cleanly. In this mixed layer the small grains pack into the spaces between the large ones and we get about as far from close hexagonal packing as it is possible to be (see p. 164). The layer therefore has a very low pore volume and a high pressure loss, and during filtration it very quickly becomes clogged so care must be taken that the two media are closely graded to the right sizes and that they stay that way. The difference in density between anthracite and sand is quite big; with other media the size grading becomes even more critical because the density differential is smaller.

7 Filter designs

We have discussed the theories of filtration in general and gone into some detail on the subject of bed filtration. This was once a very simple subject, but has become much less so in the last decade. The best way to describe the variety of bed filters and their duties is to start with the classical sand-filter design and then discuss the variations which have been developed more recently.

The previous generation of engineers only knew of 'slow' and 'rapid' sand filters. The slow filter was slow indeed: at flow rates of 0.08m/h (1–2g/ft²/h) it occupied acres of land and its action was as much biological as mechanical. It was used only by municipal works, and even there its very high specific capital cost has made it progressively less economical.

The rapid sand filter

The rapid sand filter was the only sand filter used in industry, where it continues to be the commonest design. It consists of a bed of graded sand, typically 0.5–1mm (30/15 mesh) and 600mm deep. This lies on a bed of graded gravel, in several layers, with the finest gravel at the top and the coarsest layer at the bottom. The gravel sizes go from 12–25mm, and the total depth of the gravel layers is about 300mm. Within the bottom of the gravel there is a distribution system to let water in and out.

This filter operates in downflow at rates typically between 5-10m/h (100–200g/ft²/h), and at the lower end of that range should remove all particles down to 10 microns, from waters containing up to about 100ppm of suspended matter. High flow rates are used for waters with little suspended matter and vice versa.

This means that filters are normally backwashed once or twice a day: the duty and flow rates are adjusted to one another to achieve this. Usually the filter is backwashed when the head loss across the filter rises by some predetermined value, which might be about 2.5m.

Backwashing takes a flow of about 36m/h (750g/ft^2/h) for five minutes. This means that the backwash water represents more than half-an-hour's output from the filter; this has to be filtered water, so that some thought has to be given to its supply. The waste water thus represents a loss of about 1m^3 for every 40m^3 of filtered water. Some additional means of agitating the sand is usually provided.

All these numbers are average values and are flexible over quite wide limits. Apart from that, the detail of the filter design can vary widely.

Gravity vs pressure

The head of water necessary to push water through the filter can be supplied by gravity or pressure. In municipal practice, where capital has been relatively plentiful and cheap, designers aim at low operating costs even if it means higher capital costs; filters are designed for low flow rates and low head losses. This

Plate 5.   Pressure filter installation at Pembroke oil refinery

makes it possible to use gravity filters, in which the pressure is provided by the head of water above the bed and the filter is built as an open top vessel. The same financial climate also favours concrete construction rather than steel and this, too, makes for gravity rather than pressure filters.

In modern industry, the design is motivated in the opposite direction. Industrial filters are usually pressure vessels constructed in steel, operating at higher specific flow rates and higher head losses. The filtered water from a pressure filter can be delivered under pressure and the capital cost of the re-pumping stage can thereby be avoided. The pressure vessels are always cylindrical, with the cylinder horizontal or vertical. Horizontal filters seem to have gone largely out of fashion, however.

The total throughput will of course have a great influence on the choice between concrete gravity and steel pressure filters: concrete becomes relatively cheaper for very large flows. Figures 7.1 to 7.4 show the general layouts of these kinds of filter.

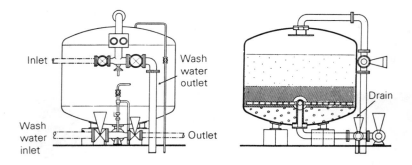

Figure 7.1. Vertical pressure filter

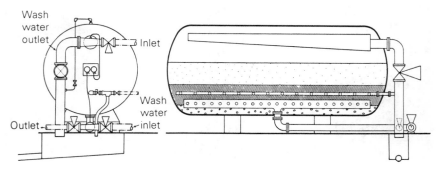

Figure 7.2. Horizontal pressure filter

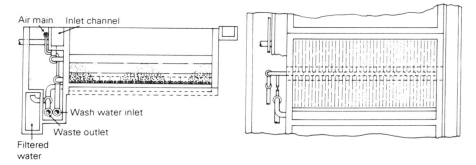

Figure 7.3. Rapid gravity filter

Backwash supply

Backwashing needs a supply of filtered water at several times the filter's rated throughput. The problem is where to take this water from and how to supply such a high flow rate. With a battery of four or more pressure filters it may be possible to use the combined output of the rest of the battery to wash one filter at a time. With gravity filters a battery can supply the volume, but a backwash pump is needed. In many cases, however, it is necessary to install a filtered water storage tank big enough to supply the necessary volume of filtered water. The output from a pressure filter can be stored in a high-level tank so that the backwash flow is then provided by gravity head. This avoids the need for a high flow rate backwash pump, an expensive piece of plant which is actually in use for only a small proportion of the time.

One highly efficient design incorporates backwash storage in the same shell as the filter itself, which greatly simplifies piping and automation. One design for an integral backwash storage filter is shown in Figure 7.4.

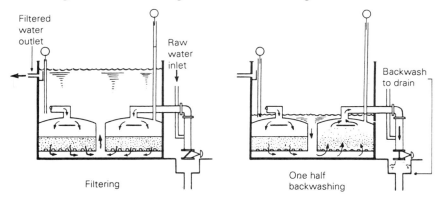

Figure 7.4. Integral backwash storage filter

Plate 6. Typical municipal filter installations (inside view)

Agitation of the bed

We have pointed out that backwash alone was insufficient to give the sand a good clean. At the conventional ratings, and in downflow, most of the dirt remains in the upper few inches of the sand, where the smallest grains lie. American practice used to be to incorporate high-pressure water jets at the top of the bed with which the sand could be properly stirred up. Other filters were built with mechanical rakes which ploughed the top of the bed. Both these devices have been largely superseded by air scouring the whole bed before the backwash proper. The agitation which this creates shakes the dirt off the sand, and the backwash is largely a flushing operation. A typical air flow is 80m/h.

Distribution

The inlet distributor at the top of the filter can be reasonably primitive because sand is a heavy material which is not easily churned up by jetting of the incoming water and because the pressure loss in the sand bed is enough to distribute the flow quite evenly.

Plate 7. Typical municipal filter installations (outside view)

The bottom distributor which is buried at the bottom of the gravel bed is classically a simple header-and-lateral systems of perforated pipes. It need not be fitted with strainer nozzles or mesh covering because the graded gravel prevents the sand reaching the distributor. On the other hand, a header-and-lateral is not efficient at distributing air, especially if this is to be done at the same time as water flows through it. Some designers overcome this by fitting a separate air distribution system. Another method is to fit special nozzles on a false bottom, which is the design shown in Figure 7.4. The nozzles are built specifically to take air and water flow simultaneously. This design costs more than the simple header-and-lateral shown in Figure 7.1, but it is more efficient.

Automation

It is good practice to terminate the filter run when the pressure loss builds up because this automatically takes into account variations in the suspended matter in the water. One early design used the rising inlet head to start a siphon which backwashed the filter, but although this scheme provided automatic operation without using any valves at all, it has not proved economical. A single valve triggered by a head sensing device can, however, be used very

Plate 8. **Integral backwash storage filtration plant and bulk chemical storage at nuclear site**

satisfactorily in filters with integral backwash storage. Figure 7.4 shows this kind of arrangement and how it works.

Filter batteries without backwash storage, or filters using separate backwash storage tanks have to be fitted with frontal pipework and valve systems.

Variations on the conventional

The standard designs described above are in common use for such duties as filtering water after coagulation, after lime softening, after oxidation (for iron removal) or just filtering of reasonably clean raw waters.

Variations on the conventional designs aim to enable the filters to handle dirtier waters or to operate at higher flow rates and so cut the capital cost. In many cases the quality of treated water is not quite so critical, which allows the designer to cut a few corners.

Upflow filters

We have said that it is clearly inefficient to let the finest sand see the heaviest suspended matter first. If the filter were run in upflow, the water would pass through progressively smaller sand grains and the filter bed would be more intensively utilized. The main problem is that upflow tends to lift the bed and this tendency increases during the run as the accumulation of dirt causes increasing pressure loss.

Very coarse media have terminal velocities so high, and cause so low a pressure loss, that they can be used in upflow without further ado. Conventional gradings, however, need something to hold the bed packed down except during washing when the bed must be free to move and expand. The Bi-Flow filter (not shown in the illustrations) has a buried collector in the upper layers of sand, through which a small proportion of the raw water flows downwards. The buried collector is a relatively expensive piece of equipment in a filter, which is after all supposed to be a simple and cheap piece of plant.

The Immedium upflow filter has a simple grid buried in the top of the bed. Bridging across the bars of the grid holds the bed during flow; the air scour breaks the bridges and allows normal washing to be carried out. This filter has proved successful for rough filtering at high flow rates of 15m/h ($300g/ft^2/h$) and even more, tackling secondary sewage effluent and similar difficult suspensions. One difficulty in upward filters is that large thin sheets of material such as tomato skins tend to wrap themselves around the inlet nozzles at the bottom of the filter. In downward filters this sort of problem cannot develop.

Upflow filters are unsuitable where a consistent quality of filtrate must be guaranteed, because they have a tendency to hiccough and allow occasional shots of dirt to break through. This may be due to occasional air bubbles rising through the bed. There is also the objection, raised especially in the context of drinking water, that the backwash goes out through the same path as the product so that there is a chance of dirt being left behind after the wash and picked up by the clean product.

The advantage of the upflow filter is the higher flow rate which reduces capital cost and the increased capacity per cycle per unit area of bed (though deeper beds are usually necessary). This increase in turn reduces the proportion of water lost in backwashing. Another important saving is that raw water can be used for washing.

A typical upflow filter layout is shown in Figure 7.5.

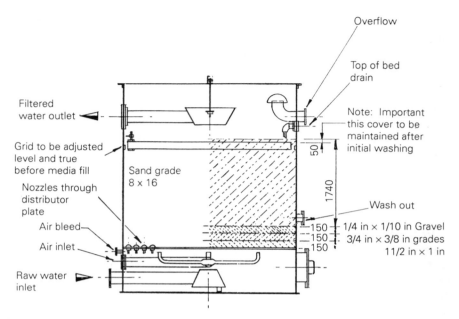

Figure 7.5. Typical upflow filter design

Very high rates and very deep beds

A conventional filter is a deep bed only in the academic sense which distinguishes it from the shallow filter layer of a precoat filter, for instance. Much deeper beds can be used.

The theoretical head loss of a clean filter bed obeys a formula in which h/l varies as v^2/d where h is the head loss, l the depth of bed, v the flow velocity and d the diameter of the grains. Coarse media will therefore permit deeper beds and higher rates to be used.

Long experience has given us the gradings in conventional filters. There is nothing to be gained by deeper beds with these. The probability of particles being held in a bed of uniform grains (by one of the mini-events described earlier in the book) is proportional to the concentration of particles at that point in the bed. The concentration of retained dirt therefore falls very quickly as we go down through the bed. In fact the conventional bed is, of course, not uniform and contains finer grains which lie at the top, which makes matters worse. In practice most of the dirt remains in the top 150mm or so of the bed. If the run is continued beyond the conventional limits of head loss, the high head loss will cause breakthrough.

Multi-media and upflow filters overcome this limitation. In the coarse medium the probability of a particle being held in any plane is reduced, and the dirt therefore penetrates deeper into the bed. In this way the coarse medium removes a fair proportion of the suspended matter without a high pressure loss, and a deeper bed can be used. If the media are so arranged that the finest medium sees only a very small concentration of suspended matter, then the filter rating can be increased, without loss of quality.

Coagulation in filters

Coagulation and flocculation as described previously (see pp. 14 – 17) are often promoted in special plant but they also take place in deep-bed filters.

The terms are used rather indiscriminately. In industrial water practice flocculation normally describes the deliberate creation of a floc, usually by dosing with alum or iron salts. Coagulation is the process by which particles stick together to form bigger particles. Others use these terms quite differently, however.

Coagulation consists of bringing small particles together, creating surfaces which will adhere to one another and avoiding shear forces which tend to rip the particles apart again.

Clearly a violent mixing action is good for bringing particles together but it is equally likely to part them again. Conditions inside a filter medium are good for coagulation. We have seen how particles tend to get trapped at the places which favour settling out; this is a more efficient way of bringing particles together than stirring the water and relying on random collisions for contact. As for the shear forces which tend to part them again, modern coagulant aids have increased the adhesion between coagulated particles to such an extent that a completely new situation has been created. A new, different, body of experience of coagulation in filters is growing as the use of deep-bed filters for this kind of duty increases. In deep-bed filters the average size of the suspended particles can be made to grow while passing through the medium. A coarse filter medium in a deep bed can therefore serve a dual function: to coagulate the particles; and to hold the bigger flocs which are of course retained more easily by the medium. This is not a new observation but it has been given fresh importance by the development of 'coagulation aids'.

These are organic compounds, soluble in water, with very long, thread-like molecules carrying electrically active groups. Suspended matter always has small charges on its surface (usually negative) and the commoner coagulant aids are therefore made to be electro-positive. The particle surfaces are therefore electrically attracted to the macro-ions (which is what the coagulant aids are).

On page 80 we describe the dissociation of ions in terms of frantic wife-swapping; here we have a more middle-aged situation. These rather stout parties do not roam about keenly but have to be brought together and introduced. The ladies don't actually like one another but will form into groups if there are a few men to hold them together.

67

Coagulant aids are expensive, but they work in very small concentrations which make the cost quite small. Their ability to hold suspended matter together in large flocs even where the water flow exerts a moderate shear force has opened a new dimension in filter practice. In allows filters to work at rates as high as 40m/h ($800g/ft^2/h$) and do a useful job. At these rates, the media have to be very coarse, otherwise the pressure loss would be too high, and the beds have to be very deep, otherwise the coarse media would not retain enough dirt. Beds of 2m (7ft) with media up to 3–4mm are not uncommon. Naturally at such high flow rates, and with such coarse media, the filters bleed a fair amount of suspended material. The usual application for such extreme designs is for rough filtration of coarse particles, such as scale in steelworks effluent, to yield a moderately clean water which can be re-used.

Continuous filters

The growing influence of chemical engineers in water treatment has led to a fashion for turning batch processes into continuous countercurrent. This fashion has included sand filters. None of the designs so far proposed is truly continuous, because tightly packed sand will not flow readily. All the designs are therefore short-cycle batch processes in which the medium is moved intermittently in such a way that the most heavily laden layers of sand are transferred to a separate washing section and then recycled to the clean end of the filter. The intention is to produce filters in which the filter beds operate at very high flow rate without breakthrough because the pressure loss in the most heavily laden end is not allowed to build up. The other advantage is that by washing only the most heavily laden medium, the water wasted in washing and the bulk of effluent can be reduced. Since both sand and water are still very cheap, the commercial prospect for continuous filters is doubtful.

Part 4
POST TREATMENT OF POTABLE WATERS

8　Post treatment of potable waters

The book so far has concentrated upon pretreatments which are applicable to waters to be used both domestically and in industry. The processes described from this point onwards have historically been mainly used by industry although this is changing. Before moving on to these subjects it is sensible to discuss the further treatment used by water undertakings whose job it is to supply drinking water.

Where drinking water is supplied from underground sources little further treatment may be required and many small private supplies are drunk with no ill effect. Where such waters contain iron and/or manganese it is normal to oxidize these by air injection to improve the flavour and reduce the scaling properties. In such cases the water is aerated and passed through a catalyst bed from which the oxidized iron and/or manganese deposit has to be backwashed off at intervals.

Surface waters are pretreated as described in the previous chapters (Chap. 2 – 4).

Before water undertakings send their water to the general public, however, they have to ensure that it is aesthetically and medically acceptable. The organic compounds in water which can give it unpleasant tastes need to be oxidized; killing the bugs requires similar treatment. The process is called disinfection and the most widely used disinfectant is chlorine. Many industial waters require similar treatment as we shall see in Chapters 22–25 but taste is not a factor in such treatments.

Even when water leaving a supplier's treatment plant is suitable for human consumption it can pick up organics/bugs from the mains, so it should have a residual disinfectant to see it through its journey.

Drinking water quality standards

The obvious standard required of all potable waters must be that they should not carry waterborne diseases or toxic substances. There is no shortage of contaminants but it would be impossible within a sensible time-frame to examine water for all possible troublemakers so routine monitoring is restricted to that of indicator organisms. If such organisms are found it does not necessarily mean that disease is present but that there is such a risk.

The EC drinking water directive allows no faecal coliforms or streptococci to be present in water at the *point of human consumption* and the WHO recommends similar standards for water *entering* the distribution systems. It also specifically recommends that water can be considered to be of suitable virological quality when a turbidity of 1NTU is achieved together with at least 0.5mg/l of free residual chlorine after a contact period of at least 30min. (This is a very brief summary.)

Chlorine

Since chlorine is the most widely used disinfectant we will concentrate on this and refer in less detail to other disinfectants.

Chlorine dissolves in water to give equal molar concentrations of hydrochloric and hypochlorous acids.

$$Cl_2 + H_2O \rightleftarrows HCl + HClO$$

The HCl dissociates into H^+ and Cl^- ions but the hypochlorous acid HClO dissociates *only partially* as H^+ and $(ClO)^-$. It is the hypochlorous acid which inactivates the micro-organisms probably by diffusion through their cells.

The equilibrium of these reactions is pH dependent such that below pH2 the chlorine is in its molecular form reforming as hypochlorous acid (HClO) at pH5 and above that the acid ionizing to complete its dissociation at pH10. Thus as the pH rises above 5 the disinfectant quality reduces.

Figure 8.1 shows this graphically and that the percentage of hypochlorous acid being maximized at pH5/6 is therefore most rapidly effective as a disinfectant at that point. As the HClO is consumed in disinfection the hypochlorite ion reassociates with hydrogen to maintain the relative proportions appropriate to the pH prevailing. Thus disinfection of a high pH water requires more time, not more chlorine.

Chlorine breakpoint

Depending upon the type of water being dosed the curve of residual chlorine

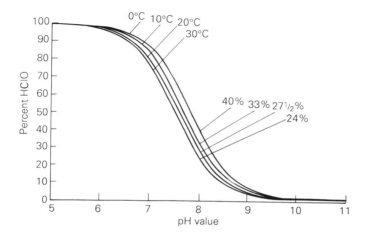

Source: Reprinted with permission from Morris, J. C. (1984), *J. Phys. Chem.* 70, 3798 Copyright American Chemical Society.

Figure 8.1. Percentage HClO in hypochlorous acid as pH varies

shows a peak and a trough as seen in Figure 8.2. This is because the chlorine reacts with the organic matter in the water and with any ammoniacal material to produce compounds which are then destroyed by further dosing; in the best regulated drinking water systems one would aim for point A which would give a small residual and lowest taste interference.

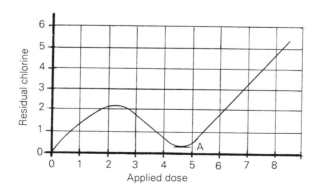

Figure 8.2. Chlorine breakpoint dosing

Dosing

Chlorine can be dosed as liquid or a gas. Several equipment supply companies give excellent service with design and installation advice including the publication of safety information. A well-engineered chlorination system will comprise:

- correctly sized equipment
- reliable metering of chlorine
- proper injection of the metered dose
- immediate, thorough mixing of the chlorine with the dosed stream
- means for adjusting the dose to keep exposure within target limits
- monitoring to check that the target exposure value is being achieved
- regular checks that the target exposure value assures bacterial quality
- appropriate maintenance procedures

A basic chlorination control scheme is shown in Figure 8.3. In such a scheme the dosing can be proportional to flow and residuals. The siting of flow meters and dosing points is critical and a time/flow logic must be set up at the design stage to ensure an optimum system. In Figure 8.3 the inlet pipe is shown with a return bend. Such designs are the result of work done to maximize plug flow in the contact tank and minimize short circuiting.

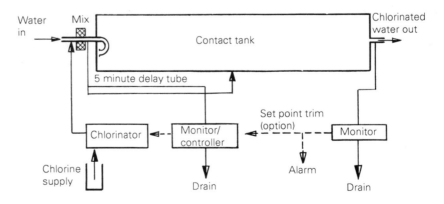

Source: Courtesy of British Effluent and Water Association.

Figure 8.3   Basic chlorination control scheme

74

Other disinfectants

Chloramines

These are chlorine derivatives which react more slowly than chlorine and therefore maintain activity for a longer period.

Chlorine dioxide

This is as powerful a disinfectant as chlorine and is useful when the water to be treated contains phenols with which chlorine would combine to give an unpleasant taste. As a pretreatment before chlorine dosing it is sometimes used to precipitate iron and manganese from borehole waters.

Sodium hypochlorite

This is a widely used disinfectant but some care is needed in its preparation as dilution can precipitate hardness from the diluting waters. It can be manufactured at site by electrolysis and there are companies which specialize in the supply of electrochlorination equipment using sodium chloride or sea water as the feedstock.

Ozone

This has the formula O_3 and is produced by subjecting a stream of clean dry air to a very high voltage. The gas produced is, as might be expected, rather unstable. The process generates considerable heat which reduces the O_3 yield so the system has to be cooled. Ozone generation is a specialist technology with specialized equipment suppliers. When giving consideration to the use of ozone it should be borne in mind that further dosing of chlorine to achieve lasting disinfectant residuals may be necessary.

Ultraviolet

A useful disinfectant at the point of use is an ultraviolet lamp. This produces a radiation which kills bugs but gives no residual protection. Manufacturers of such equipment are continually improving their product, partly in response to its increased use in the belt and braces world of ultra-pure water production.

From the foregoing it will be seen that the potable water supplier has available a wide range of disinfectant and oxidizing processes for the protection of the consumer. Most of these are associated with well-designed and proven equipments available from specialist manufacturers.

Inorganics

The ability of scientists to measure and monitor the effect of such constituents of water as nitrate, aluminium and sodium is leading to great debate and massive expenditure in developed countries. A topical example is the decision to limit the nitrate in potable water to about 50ppm by the EC. In farming areas where crop enhancement has been achieved by the use of nitrogen as fertilizer, high nitrogen levels in the soil lead to nitrate leaching into water supplies. Many of these supplies will now have to be treated to reduce their nitrate levels. Chaper 18 is devoted to achieving this by ion exchange.

Aluminium is said to be a factor in the development of Alzheimer's disease. Since many surface water sources have a relatively high aluminium background and these same waters are flocculated for settlement using aluminium sulphate the problem is compounded and methods of pretreatment may have to be changed.

For many years scientists have told the consumer that there is some correlation between sodium levels in drinking water and heart disease. It is for this reason that the British Effluent and Water Association recommends that domestic water softeners bypass the drinking water tap. It is not so many years ago that municipal water authorities were installing softeners in mains supply routes.

Science will find plenty yet to worry us about the water we drink and the treatments required to remove these worries will continue to blur the definitions of what is domestic and what is industrial water treatment.

Part 5
ION EXCHANGE AND DEMINERALIZATION

9   Ion exchange

The behaviour of ions

Water has some most unusual properties when compared with other fluids. Engineers are familiar with its high latent heat which means that a kilogram of steam carries more energy than any other vaporized liquid. Its main interest to chemists lies in its excellent properties as a solvent and the fact that it is a powerful 'polar' solvent, which means that substances dissolved in it ionize very readily.

Ionization

The commoner salts such NaCl dissolve in water as molecules but, as soon as the molecules are in solution, the polar action of the water makes them split up. In the case of NaCl, this split is into Na^+ and Cl^- ions. An ion is an atom or group of atoms with an electrical charge, positive or negative (shown by the sign in the formula). Some ions have more than one charge; for example magnesium sulphate ionizes and the $MgSO_4$ splits into Mg^{++} and SO_4^{--}. Another example is calcium chloride, $CaCl_2$ which gives one ion of Ca^{++} and two of Cl^-. In every case the sum of all the charges must of course be zero, otherwise one would get an electrical shock from the solution. As long as the solution is dilute, the common inorganic salts dissociate completely and for our purposes exist only in the form of ions.

A wife-swapping party

Once a salt has dissociated into ions each ion is absolutely free to move about and form combinations with other substances. Ions do not 'remember' how they come to be in a solution. For example, a solution can be made with sodium chloride (NaCl) and potassium nitrate (KNO_3) or with potassium chloride (KCl) and sodium nitrate ($NaNO_3$). If the right quantities are used, the two solutions can be made absolutely identical.

One way of thinking of salt solutions is as a wife-swapping party to which the Smiths and Joneses arrive, tidily enough, as married couples. As soon as they're in the party, they immediately part company and roam about singly (and highly charged) looking for some congenial partner and without paying the faintest attention to their original spouses. The comparison is frivolous, but it will serve quite well, as we shall see.

Dissociation of water and conductivity

Water itself dissociates, but only to a minute extent. The H_2O molecule splits into H^+ and OH^- ions, but there is a rule that the concentration of these two ions multiplied together must always by 10^{-14}. (Table 9.2 will show appropriate units.)

Ions normally roam about a solution quite at random, equally in all directions. However, by moving in an orderly direction, they can carry an electrical current. Water itself dissociates to such a minute extent that there are very few ions to carry current. Pure water is therefore a very poor conductor indeed but dissolved salts create more ions which increase its conductivity, so that a conductivity measurement is a good indication of the quantity of ions in solution.

Acids and alkalis

Acids and alkalis can be thought of as special kinds of salts; all acids dissociate to yield H^+ as the cation, but ordinary anions similar to those of normal salts; all alkalis yield ordinary cations but OH^- as anions. The H^+ and OH^- ions each have strong and characteristic properties, and it is these which make acids and alkalis behave in their characteristic manner.

We have already seen that solutions of ions in water have to obey two rules. Rule 1: the total number of cations and anions (measured by their electrical charges) must always be equal to one another. Rule 2: the H^+ concentration multiplied by the OH^- concentration must always equal 10^{-14}.

If H^+ or OH^- ions are introduced into the water from some other source, then to meet the second rule the dissociation of the water itself is reduced to offset the increase in the product of the two ions multiplied together.

H^+ and OH^- ions

This has quite interesting consequences which can be shown if we imagine a quantity of water as illustrated by the examples in Table 9.1.

Table 9.1. Dissociation examples

Example	Values as g equiv/l	CATIONS				ANIONS				$H^+ \times OH^-$
		H^+ (from water)	H^+ (other)	Na^+	Total cations	OH^- (from water)	OH^- (other)	Cl^-	Total anions	
1	Pure Water	$1/10^7$			$1/10^7$	$1/10^7$			$1/10^7$	$1/10^{14}$
2	Pure Water + $10/10^7$ NaCl	$1/10^7$		$10/10^7$	$11/10^7$	$1/10^7$		$10/10^7$	$11/10^7$	$1/10^{14}$
3	Pure Water + $10/10^7$ HCl	$0.1/10^7$	$10/10^7$		$10.1/10^7$	$0.1/10^7$		$10/10^7$	$10.1/10^7$	$1/10^{14}$
4	Pure Water + $10/10^7$ NaOH	$0.1/10^7$		$10/10^7$	$10.1/10^7$	$0.1/10^7$	$10/10^7$		$10.1/10^7$	$1/10^{14}$

Example 1 If the water is initially pure it contains 10^{-7} of H^+ and 10^{-7} of OH^-. The concentration of cations equals that of anions and their product equals 10^{-14}. Both rules are satisfied.

Example 2 If a small amount, say 10^{-6} of NaCl, is dissolved in the water it dissociates to form equal concentrations of Na^+ and Cl^-. These add themselves to the ions already present in the water and the two rules are still satisfied. The balance of H^+ and OH^- ions has not been altered.

Example 3 However, if we add some HCl, say 10^{-6}, being a strong electrolyte it splits entirely into H^+ and Cl^- ions. The resulting ionic concentrations that we might expect arithmetically from this would be:

$$H^+ \quad 1.1 \times 10^{-6}$$
$$OH^- \quad \times 10^{-7}$$
$$Cl^- \quad \times 10^{-6}$$

The number of cations is equal to the number of anions, obeying Rule 1, but the product of H^+ and OH^- concentrations would be 1.1×10^{-13} (that is 1.1 $\times (10^{-6}) \times (10^{-7})$) which breaks Rule 2. In fact this condition would not exist:

most of the H^+ and OH^- ions which were present from the pure water would recombine to form H_2O molecules ensuring that Rule 2 is satisfied.

We can calculate quite closely how much of the water recombines. To meet Rule 2 we have to satisfy the equation

$$(10^{-6} + H^+) \times (OH^-) = 10^{-14}$$

where the 10^{-6} represents the H^+ ions from the HCl, and the H^+ and OH^- represent the ions from dissociated water. But to satisfy Rule 1, the H^+ and OH^- must be equal, so we can say that

$$H^+ = OH^- = W$$

Then
$$(10^{-6} + W) \times W = 10^{-14}$$

or
$$10^{-6} W + W^2 = 10^{-14}$$

Obviously W is going to be small so we can neglect W^2 for a reasonable approximation and then

$$10^{-6} W = 10^{-14}$$

or
$$W = 10^{-8}$$

This is a very close approximation of the concentration of OH^- ions in the solution, thereby satisfying Rule 2: that is $10^{-6} (H^+) \times 10^{-8}(OH^-) = 10^{-14}$. Putting this result into words, by introducing HCl we have forced so many H^+ ions into solution that most of the H^+ and OH^- ions from the pure water recombine to form H_2O molecules and maintain equilibrium.

Example 4 In this case we have added an amount of NaOH equal to the concentration of 10^{-6} to the pure water. This time the effect is to create an excess of OH^- ions, and by following similar arithmetic to that used for Example 3 we find that it is now the H^+ concentration which has fallen to 10^{-8}.

The meaning of pH

Any aqueous solution has some H^+ ions in it. If we measure their concentration and write it in the form 10^{-x}, then x represents the 'pH index'. In Example 1 we have pure water with 10^{-7} H^+ ions in it, so the pH is 7. In Example 3 we have 10^{-6} H^+ ions, so that the pH value is 6. In Example 4 we have 10^{-8} H^+ ions, giving a pH value of 8. This is how the pH value measures acidity and alkalinity.

The actual unit of measurement used by physical chemists for this calculation is the gram equivalent/litre, which is 1000meq/l. Table 9.2 shows the concentrations which this really represents. The important points are: the concentration of H^+ and OH^- in near-neutral waters is extremely small; and every unit of pH represents a tenfold change of concentration. The acidity at pH2 is 1000 times that at pH5.

Table 9.2. The concentrations represented by different pH values

pH value	Hydrogen ion concentration			OH^- ion concentration ppm as $CaCO_3$	Equivalent dose in pure water	
	as g equiv/l	as meq/l	as ppm $CaCO_3$		HCl ppm as such	NaOH ppm as such
2.0	$\frac{1}{10}^2$	10.0	500.0		400 approx	
3.0	$\frac{1}{10}^3$	1.0	50.0		36	
4.0	$\frac{1}{10}^4$	0.1	5.0		3.6	
5.0	$\frac{1}{10}^5$	0.01	0.5	↑ etc	0.36	
6.0	$\frac{1}{10}^6$	0.001	0.05	0.0005	0.03	
7.0	$\frac{1}{10}^7$	0.0001	0.005	0.005	nil	nil
8.0	$\frac{1}{10}^8$	↓ etc	0.0005	0.05		0.04
9.0	$\frac{1}{10}^9$		↓ etc	0.5		0.4
10.0	$\frac{1}{10}^{10}$			5.0		4.0
11.0	$\frac{1}{10}^{11}$			50.0		40.0
12.0	$\frac{1}{10}^{12}$			500.0		450 approx

Natural waters

Natural waters are dilute solutions, in other words the salts they contain are completely dissociated. It is actually meaningless to talk of a water containing, say, calcium bicarbonate, but we must think of a water containing (amongst others) Ca^{++} and HCO_3^- ions.

The ions most commonly found in waters are:

Cations	Anions
Calcium Ca^{++}	Bicarbonate HCO_3^-
Magnesium Mg^{++}	Chloride Cl^-
Sodium Na^+	Sulphate SO_4^{--}

Every natural water contains all these in some proportion as well as the inevitable traces of H^+ and OH^- ions; the whole mixture still has to meet the two rules which we have discussed above. However, natural waters usually have a pH so near to 7 that the concentrations of H^+ and OH^- ions can be neglected (see Table 9.2). The sum of the cations must therefore equal the sum of the anions shown in the analysis.

Bicarbonate — the joker in the pack

Of these ions, the bicarbonate ion HCO_3^- has the most peculiar properties. It is formed from dissolved CO_2 in water, a process which is reversible. In the cold, and at pH above 4, a part of any CO_2 dissolved in water combines with an H_2O molecule like this:

$$CO_2 + H_2O \rightarrow H^+ + HCO_3^-$$

This creates H^+ ions, and therefore lowers the pH. At pH4, the process stops, and any surplus CO_2 just remains in the water as a dissolved gas. By dissolving enough CO_2 in any water, however, its pH can always be lowered to 4. Below pH4, or in the hot, the same reaction goes backwards:

$$HCO_3^- + H^+ \rightarrow CO_2 + H_2O$$

Now the reaction removes H^+ ions from the solution and therefore raises the pH. This can have nasty effects, such as the formation of calcium or magnesium scale in boilers. Another way of pushing the reaction in this direction is to remove the dissolved CO_2 from the solution, such as by boiling it or blowing it with air. If this is done to a solution whose low pH is due to CO_2 alone, the pH can be raised to very nearly 7. But CO_2 which is boiled out of boiler water in this way may well redissolve in cool condensate somewhere else in the boiler system and reduce its pH, possibly to as low as pH4. This is a very corrosive solution, hence the danger of bicarbonate and CO_2 in boiler feed water.

Ion exchangers

Ions are free to move about in water (in fact the name is Greek for wanderer). However, it is possible to make chemical structures in which ions are bolted on to a framework. An ion exchange resin is a plastic lattice-work with ions fixed very firmly on to it. Each of these fixed ions must have a counter-ion of opposite charge loosely attached to it, but the loose ion can be exchanged for another one if the conditions are right.

For example, a modern cation exchanger consists of a polystyrene matrix with SO_3^- groups fixed to it. Each SO_3^- group must have a cation, for example Na^+, loosely associated with it but the Na^+ (or whatever ion it is) can quite easily be exchanged for another cation, hence the name cation exchanger. In the following chapters we shall see how this property is put to use.

10 Weak and strong ion exchange resins

We have described ions in terms of sex, as two opposite kinds of creature in search of an attractive partner. Not all ions are like this; like people, some can take it or leave it. Given an attractive partner they get quite keen but when there are none they lead a blameless and inactive existence.

The HCO_3^- ion is like this and we have described how, given a supply of Na^+ or Ca^{++} ions, it dissociates just like the Cl^- and other lusty anions. But when these ions are absent and H^+ ions appear for electrical neutrality, that is, when pH is low, then the HCO_3^- loses its charges, pairs off with an H^+ and reverts to H_2CO_3 which usually decomposes to form H_2O and CO_2.

Ion exchangers (such as ion exchange resins), which are chemical frameworks with ions bolted on to them, can carry active groups of this kind, either anionic or cationic, and we will now deal with weak cation and anion resins and see how they compare with strong resins.

Strong and weak groups

A strong group (whether free or fixed on to an ion exchanger) is dissociated and active at any pH; a weak group only dissociates when counter-ions are freely available. The HCO_3^- ion goes out of business below pH4, and so does the $-COO-$ group of a weakly acidic resin (see Chapter 12). Below pH4, the group takes a H^+ ion and becomes $-COOH$, and in this form the H atom is fixed and not exchangeable. As far as the weakly acidic resin is concerned, the party is over.

A group with so little interest is easily robbed of any partner it may have acquired. When a resin is in the $-COO-Na^+$ form and the water around it falls below pH4, the resin immediately gives up the Na^+ and reverts to $-COOH$. This means that weak groups are very easily regenerated.

Weak anion exchangers behave just like this only at the other end of the pH scale. They do not dissociate at high pH, but are easily regenerated with small excesses of caustic soda or even weaker bases such as ammonia, soda ash or lime water.

Chemical efficiency

This term deals with the amount of regenerant used rather than the degree of ion removal, which is termed 'slip' or 'leakage'. The best way of measuring it is as a regeneration ratio, which is the regenerant put on to the resin divided by the capacity resulting, both measured in gram equivalents, pounds as $CaCO_3$, or some other unit of chemical equivalence.

For example, a litre of cation resin is regenerated with 1.60g equiv of H_2SO_4; 1.05g equiv of acid passes through unchanged and 0.55g equiv of H^+ ion remains on the resin and is available as the capacity of the next run. Then the regeneration ratio is

$$\frac{1.60}{0.55} = 2.91$$

This set of numbers is the same as: a cation resin is regenerated with 5lb/ft^3 of H_2SO_4, giving a capacity of 11.5kgrn/ft^3 as $CaCO_3$ (see p.343). Obviously calculation with decimal units is much easier.

Other ways of describing these figures would be to say that the efficiency of regeneration is 34 per cent, that an excess of 191 per cent of acid is used or that the acid used is 291 per cent of theory. But percentages are misleading unless the basis is quite clear. Also, the regeneration ratio is a handy number and can be used directly in the arithmetic.

Regeneration ratios are not easily measured on full-scale plant. One can usually, but not always, measure the regenerant put on the resin but it is generally difficult to measure directly how much passes through unchanged. The resin capacity is calculated from the length of the next run but since any one run may not fully exhaust the resin, a single reading cannot give a reliable answer. It is best to take readings over three successive cycles.

Chemical efficiency of cation exchangers

The example above is fairly typical of strong cation performance. These resins hold Ca^{++} so strongly that using H_2SO_4 high regenerant ratios are needed to knock it off. Na^+ is less strongly held, but even with Na^+ a ratio less than 1.5 is impractical. Ca^{++} predominates in normal waters so that ratios of 2–3 must be used to obtain a reasonable capacity and leakage. Low capacities at low ratios do save chemicals but lead to the use of large volumes of expensive resin, which cancels the saving.

The same arguments hold for strong anion exchangers.

Chemical efficiency of carboxylic exchangers

We have said that the weakly carboxylic group goes out of dissociation below pH4 and becomes $-COOH$. In order to do so it gives up any Ca^{++} or Na^+ it may hold, and once it has done so it does not show any desire to take them back as long as the pH remains low. The affinity differences which it showed when dissociated do not enter into this reaction. The regeneration therefore proceeds as long as the resin is in water below pH4. Acid is used to supply H^+ ions equivalent to the cations which leave the resin, and the only excess which is necessary is the amount required to neutralize the alkalinity of the raw water used for regeneration and rinse. The regeneration ratio depends partly on the alkalinity of the raw water, and is normally below 1.1.

In fact the resins are so easily regenerated that within a wide band the capacity depends largely on the amount of acid put on the resin. Weak resins are not as quick acting as strong resins, which limits the specific flow rate at which they can be used to 40BV/h ($4g/ft^3/min$). (BV stands for bed volume, a very handy concept indeed. It refers to the gross settled volume of the bed and is not to be confused with the void volume (VV) which is about 40 per cent of the BV. The charm of the BV is that it is a dimensionless ratio, equally useful with English or decimal units.)

A good carboxylic resin normally gives a capacity of 1.2–1.8g equiv/1 (4–6lb $CaCO_3/ft^3$). This high capacity combined with a limited specific flow rate can be quite an embarrassment to the designer. He is forced to use a minimum volume of resin (which is not cheap) which will give him a longer run than he may need, at the end of which it uses a large bulk of regenerant which calls for large regeneration equipment.

The ease of regeneration leads to other uses for carboxylic resins. For example, demineralization plant effluents are acid and alkaline in turn. The plant can be designed to give a self-neutralising effluent but to mix the dilute effluents needs a very large tank. It may be cheaper to pass the effluent through a weakly acidic unit acting as a buffer. When the effluent is alkaline, the resin takes up Na^+ ions and gives water between pH4 and 6. The acid effluent which follows takes the Na^+ off again and comes out at pH4. The usual way to control these systems is to have a slight excess of alkali, and if the unit has enough capacity to accommodate the swing, the effluent stays within a narrow, near-neutral band.

Strong and weak base resins

Theoretically these are the same as cation resins, only opposite. Weak base resins exchange anions but only in a low pH environment. They, too, have a high working capacity because of their ease of regeneration, with ratios between 1.3 and 1.6 being normal. They, too have to be used below 40BV/h (4g/ft^3/min) because of their slow reaction rate. Their normal capacity is 0.9–1.2g equiv/l (3–4lb CaCO$_3$/ft^3).

Strong base resins, on the other hand, are normally regenerated at ratios of 3 or more in order to obtain a reasonable capacity and to keep the silica leakage down. Even then their normal capacity is less than half that of the usual weak base resin capacity.

Weak base resin behaviour in practice

The water coming from a strongly acidic unit contains little besides free mineral acidity (FMA), which a weak base resin will remove very efficiently.

Two factors confuse this simple picture. First of all, the distinction between strong and weak base resins is not black-and-white but strong base resins contain a few weak groups and vice versa. This partly accounts for the higher regeneration ratios which the weak base resins need (as compared with weakly acidic resins).

The other point is that water from cation units contains, in addition to FMA, weak acids which cannot maintain a low pH. The one which really matters in this context is bicarbonate, which occurs to some extent in all natural waters.

Suppose, for example, that we are to treat London water, and we have a weak base unit after the cation unit. The raw water contains:

Cations		Anions		Non-dissociated	
Hardness	314	Alkalinity	229	CO$_2$	45
Sodium	94	EMA	179	Silica	12
Total	408	Total	408		
(all as ppm CaCO$_3$)					

After the cation unit, this will be converted to:

Cations		Anions*		Non-dissociated	
Na$^+$ (say)	10	FMA	169	CO$_2$	503**
H$^+$	169	EMA	10	Silica	12
Total	179	Total	179		

The 169ppm of FMA which the water from the cation unit contains will give it a pH of about 2.5. When this gets on to the regenerated weak base resin, the top of the bed keenly takes up Cl^- and SO_4^{--} ions. As a result, the pH rises, and as soon as it gets to pH3.8, the CO_2 in the water begins to revert to HCO_3^- ions, and these get taken up by the resin too. The pH goes on rising, and at pH7 the resin stops dissociating, but by the time the pH reaches this level almost all the HCO_3^- ions have been taken up on to it. In practice there are always a few strong base groups on the resin so that at first the pH actually goes higher than this and the first runnings from a weak base unit may be as high as pH10.

The next lot of water which gets on to the top of the weak base bed therefore finds resin loaded with Cl^-, SO_4^{--} and HCO_3^- and unable to take on any more. The pH therefore stays at 2.5. Bicarbonate cannot exist at this pH and any HCO_3^- ion which strays out of the resin immediately reverts to CO_2. This sets up a one-way traffic whereby Cl^- and SO_4^{--} ions from the water displace HCO_3^- ions on the resin while the water is enriched in CO_2. When this lot of water gets lower down the bed, however, all the CO_2 is taken on to the resin again only to be displaced in turn. In this way, a CO_2-rich front of water moves down the resin bed.

This means that the very first runnings from the bed at pH9 or 10 are followed by a period during which the water is at about pH7 and contains no CO_2. Some time in mid-run, a high concentration of CO_2 breaks through, much higher than that in the water entering the unit. At this point the pH is about 5. The run then continues until FMA breaks through at pH4 and the resin has to be regenerated.

Silica on the other hand forms an anion which is so weak that the few strong base groups on the resin cannot hold it, and the raw water silica goes through the whole run at the same concentration.

This wave of CO_2/HCO_3^- is actually necessary for some weak base resin types. These are resins which do not take up Cl^- very easily; taking up HCO_3^- pre-swells them and allows the Cl^- to get on more easily. Without CO_2 these resins do very badly on waters which contain mostly Cl^-.

---

* The anions which are capable of forming strong acids with H^+ ions are called the equivalent mineral acidity or EMA. If water containing EMA is passed over a cation exchanger in the H^+ form, then this latent acidity becomes actual free mineral acidity or FMA. The ions concerned are mostly SO_4^-, NO_3^- and Cl^-.
** The amount of CO_2 produced by bicarbonate. A lot of confusion can arise when ionic concentrations and CO_2 are measured as 'ppm as $CaCO_3$'. The point to hang on to is that the ions and molecules of CO_2, $CaCO_3$, HCO_3^- and CO_3^{--} each contain one atom of carbon. When they change into one another, any one of them can only become one of the others, but starting with CO_2 there is the possibility that it will go to monovalent HCO_3^- or divalent CO_3^{--}. The convention of measuring 'as $CaCO_3$' assumes that it will go to divalent CO_3^{--}, though in HCO_3^- it has only gone half way. Hence the rule that *1ppm of bicarbonate as $CaCO_3$ produces 2ppm of CO_2 as $CaCO_3$.*

Practical problems with weakly acidic resins

The modern weakly acidic resin, based on polyacrylics, has a very high capacity indeed. When it takes up a large ion like Na^+ in exchange for a small one like H^+, the resin swells and if a large part of its total capacity is used then the volume of resin can increase by 50 per cent. In normal water treatment, when less than half the total capacity is used, the resin swells by about 10 per cent during the run. At the same time the resin is kept compressed by the downflow of water, and these resins therefore tend to agglomerate, especially if the pressure loss across the bed is high. Designers have to watch out for this, and users have to backwash the bed carefully to break up any tendency to 'graping' of the beads.

The resin's slow reaction rate limits the specific flow rate to $40BV/h$ ($4g/ft^3/$ min), but even below this limit the usable capacity varies quite considerably with the flow rate and also with the temperature, because higher temperature speeds the reaction and so increases the capacity.

H_2SO_4 regeneration of cation resins results in $CaSO_4$ in the spent regenerant. This has a solubility of only 0.15 per cent, so that there is a danger of its precipitating in the unit. If it does so, it can clog it completely, or if the precipitation is slight the crystals which remain behind redissolve slowly during the next run and bleed Ca^{++} and SO_4^{--} into the treated water.

Fortunately the precipitation is rather slow so that, by pumping the acid in fast, it is possible to run at super-saturated concentrations without the $CaSO_4$ actually coming out within the unit: some operators run it so fine that the effluent can be seen turning milky in the drain.

This problem is common to all cation units treating hard waters and regenerated with H_2SO_4 but it is more acute with weakly acidic resins. First, the normal duty for these resins is removal of temporary hardness, so that the proportion of Ca^{++} in the ions taken up is high. Second, the solubility of $CaSO_4$ is higher in the presence of excess acid, but the weakly acid resins are so efficient in regeneration that very little free acid gets through the unit and that only at the very end of the regeneration.

In normal practice, therefore, weakly acidic resins must be regenerated with H_2SO_4 at 1 per cent or weaker, and this means that the regeneration takes a lot of time and waste water. The use of HCl avoids this problem, but if a strong solution is pumped in quickly the sudden shrinkage of the bed can cause trouble.

Practical problems with weak base resins

Once again the specific flow rate is limited to $40BV/h$ ($4g/ft^3/min$). Below this limit, again, the capacity is dependent on the flow rate, but these resins are less temperature-sensitive.

Some early weak base resin types oxidized slowly in use and as a result acquired carboxylic $-COOH$ groups. This is a weakly acidic group and behaves exactly as you would expect: the high pH during regeneration with caustic soda makes it pick up Na^+, and when low-pH water runs over it during the run, the Na^+ is regenerated off again. In its milder form, this results in an excessively long rinse until all the Na^+ is displaced, but sometimes it gets so bad that a bleed of sodium persists for the whole run.

Modern resins are more stable, but while this fault is now avoided, organic fouling leads to exactly the same symptoms.

Most organic matter in water is weakly acidic, especially in mountain waters which are very rich in humic and fulvic acids from decaying plant matter. These organics tend to stick to the resin and then their weakly acidic groups result in the same symptom of long rinses and sodium bleed. Special resins can be used to ease the problem but do not necessarily cure it.

The obvious place in the flow sheet for a weak base unit is just after the cation unit and before the degasser. The weak base unit raises the pH of the highly corrosive cation water so that the degasser and degassed water pumps can then be made of cheaper materials. If the resin is a type which needs CO_2 to pre-swell it, this flow sheet is essential. The degasser after a weak base unit must be designed to cope with the wave of high CO_2 which comes out in mid-run but this is not normally of much import in degasser design.

Weak resins in ion exchange trains

The cheapest demineralization plant is a strong cation and anion unit, or just a mixed bed. Though cheap to buy, such a plant uses a lot of chemicals. It is only economical where the annual ionic load is low. (The annual ionic load is the weight of $CaCO_3$ (in pounds or g equiv) to be removed per year, and is therefore proportional to the flow rate, the annual utilization and the total dissolved solids or TDS.)

For high annual ionic loads, more complicated plants reduce the chemical running cost. The first saving can be made with a degasser tower which reduces the anion load by removing the alkalinity, and the high cost of caustic soda makes this particularly attractive.

Further savings can be made by using weak cation and anion units which remove part of the TDS at much better regeneration ratios. The combinations and permutations depend on the annual ionic load, the raw water analysis and the economic parameters which are applied to the design. We will discuss these variables and their effects on pages 99 – 106.

11   Base exchange (softening)

Base exchange means the exchange of the cations which cause hardness, Ca^{++} and Mg^{++}, for Na^+ ions which are relatively harmless to low pressure boilers. It is the oldest and simplest application of ion exchange. Even so, there are many users of the process who only understand superficially how a base exchange unit really works.

Base exchange uses cation exchangers. These are synthetic resins which have negatively charged groups fixed on to the resin matrix, and each of these groups has to have a positively charged ion loosely attached to it so that the resin is always electrically neutral. These loosely attached cations can be exchanged quite easily for other ions which happen to be in the solution surrounding the exchanger and, as the resin is porous, the solution is able to penetrate right into the resin and react with the active groups in the middle of the resin particles as well as those on the outside.

Affinity differences

All ion exchange operations make use of the fact that the resin would rather be associated with some ions than others. As a general rule, a resin has a high affinity for ions with a large number of electrical charges, that is, it prefers Ca^{++} with its two charges to Na^+ which has only one. Between ions with the same number of charges, the resin prefers the bigger and heavier ion, and so for instance it prefers Ca^{++} (equivalent weight 20) to Mg^{++} (equivalent weight 12) and it prefers Na^+ (equivalent weight 23) to H^+ (equivalent weight 1).

There is therefore a pecking order of affinities, in which the commoner cations can be listed like this:

$$Mn^{++++} > Fe^{++} >> Ca^{++} > Mg^{++} >> Na^+ >> H^+$$

The list distinguishes between large affinity differences and small ones. For example, the affinity difference between Fe^{+++} and Ca^{++} is large because Fe^{+++} has one more charge, while the difference between Ca^{++} and Mg^{++} is small because they have the same charge and the resin only chooses between them by virtue of a small difference in their weight. By contrast, the affinity difference between Na^+ and H^+ is large, because of their very large weight difference.

These affinity differences are greatest in dilute solution and they become relatively small in strong solution, though the pecking order generally remains unchanged.

Earlier, in Chapter 9 it was suggested that an ionized solution was like a wife-swapping party to which only couples were admitted, so that the number of men and women was always equal, and in which the couples split utterly, each partner becoming entirely independent.

Seen in this light, an ion exchange resin might be considered as a bench with girls chained on to it; each of the girls must always be accompanied by a man, but they prefer some kinds of men to others.

Equilibrium

When resin and water come into contact with one another, exchange of ions between the resin and the solution is constantly taking place. If the solution contains, say, Ca^{++} ions and the resin is predominantly in the Na^+ form, then the preference which the resin has for Ca^{++} means that it will tend to take up Ca^{++} and allow Na^+ out into the solution. However, the Na^+ in the solution also makes efforts to get back on the resin, and a very small proportion of Na^+ ions actually succeeds. Finally an equilibrium is set up, that is, a state of affairs when the number of Ca^{++} ions going on to the resin is just the same as the number of Ca^{++} ions being pushed off by Na^+ ions. The net result of this two-way movement is zero, and so the reaction appears to have stopped.

In Figure 11.1 we show a dilute solution of Ca^{++} ions (such as might be produced by dissolving $CaCl_2$) being brought into contact with a resin which is entirely in the Na^+ form. The resin has a much greater affinity for Ca^{++} and so at first the exchange goes violently towards getting all the Ca^{++} on to the resin. As this exchange proceeds, however, the concentration of Na^+ ions in the solution must increase correspondingly, for the total concentration of ions in the solution remains unchanged. The rate at which Na^+ ions successfully work their way back on to the resin increases with the number of Na^+ ions in the solution. Finally there are only a very few Ca^{++} ions in the solution competing with a large majority of Na^+ ions and at this point equilibrium is reached.

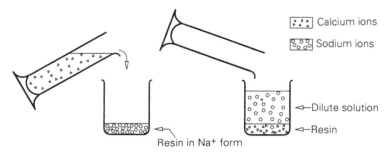

Figure 11.1.   Resin affinity for calcium ions

If we now drain the water off this sample of resin, and replace it with a strong NaCl solution, the difference in the resin's affinity for the two cations is reduced: it still prefers Ca^{++} to Na^+ but the Na^+ becomes relatively more competitive. Figure 11.2 shows this new condition when a new equilibrium is reached. Because of the reduction in the affinity difference, it is possible to get perhaps one third of the total active groups on the resin back into the Na^+ form, while the resin is in contact with a solution which is still two-thirds Na^+ ions.

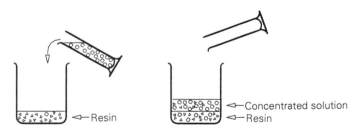

Figure 11.2.   Resin regeneration by sodium ions

Column operation

These two examples represent the exhaustion and regeneration stages of the ion exchange cycle. In practice, however, the water and regenerant are not stirred with the resin but flow through a column of settled resin. The equilibrium which each individual bead of resin reaches is governed by the same rules but column operation introduces some complications of its own. In the first place, natural waters contain both Ca^{++} and Mg^{++}, as well as some Na^+. Also, a finite time is needed to establish equilibrium between resin and liquid and, with water flowing fast down a column, this cannot always be achieved.

Suppose we have a column of resin entirely in the Na^+ form down which we pass a natural well water having, say, the following cations:

$$
\begin{array}{lll}
Ca^{++} & 225\text{ppm as } CaCO_3 \\
Mg^{++} & 50\text{ppm as } CaCO_3 \\
Na^+ & 25\text{ppm as } CaCO_3 \\
\\
Total & 300\text{ppm as } CaCO_3
\end{array}
$$

When the first drop of water contacts the first bead at the top of the column, Ca^{++} and Mg^{++} are rapidly taken up by the resin and the first drop goes on down the column as an almost totally Na^+ solution. Passing on through an all Na^+ resin, this drop is in equilibrium with the rest of the column so no effective exchange takes place below the first bead and a softened water emerges from the bottom of the unit.

This first drop of water left behind Ca^{++} and Mg^{++} on the first bead, in the proportion in which they occur in the raw water. The next drop of raw water now comes into contact with the same first bead. The affinity for Ca^{++} is greater than that for Mg^{++}, and so the resin takes up some more Ca^{++} in exchange for which it rejects Mg^{++} into the water. The resin therefore ends up with a higher proportion of Ca^{++} than the raw water, while the second drop which passes on down the unit contains a high proportion of Mg^{++}. This Mg^{++} enriched water then arrives at the second bead, which is still in the all-Na^+ form and therefore takes up all divalent ions keenly while rejecting Na^+ ions.

The third drop of water finds the first resin bead containing a preponderance of Ca^{++} and therefore in equilibrium with it. It finds the second bead enriched with Mg^{++}, which it pushes off down the column.

The ion exchange zone

In this way an ion exchange zone is built up, which has the following characteristics:

- Above the ion exchange zone, the incoming water and the exhausted resin are in equilibrium with one another, so that raw water passes through it unchanged.
- Below the ion exchange zone, the treated water and regenerated resin

are in equilibrium with one another so that the water passes through it unchanged also. All exchange takes place within the zone.

- As water flows down the column and exchange proceeds, the zone moves slowly down the column.
- Inside the ion exchange zone, there is a tendency for less tightly held ions like Mg^{++} to be pushed in front, creating a front of water enriched in Mg^{++} which runs ahead of the raw water.
- This front of Mg^{++} rich water creates a band of Mg^{++} rich resin which moves down the column at the head of the exchange zone.
- At the beginning of the run, the exchange zone is very shallow, but it deepens as it moves down the column.
- When the head of the ion exchange zone reaches the bottom of the unit, the resin bed is normally considered to be exhausted and the unit is taken out of service for regeneration. At this point the breakthrough of hardness is mostly of Mg^{++} ions, because it is the enriched front which is breaking through.
- At this point, only the resin above the exchange zone is totally exhausted: within the exchange zone there is an Mg^{++} rich zone which could remove Ca^{++} in exchange for Mg^{++}, and at the very bottom there is also a small amount of Na^+ which has not yet had time to come to equilibrium with the advancing front.

This property of taking up ions in bands according to their affinity is called chromatography and is used in some chemical processes for the separation of, for example, rare metals.

Figure 11.3 shows how the ion exchange zone moves down a column during the exhaustion stage of the cycle. Figure 11.4 shows the concentration of different ions in the resin and in the water within the ion exchange zone.

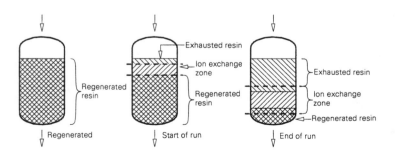

Figure 11.3. Progress of ion exchange zone

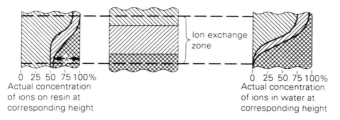

0 25 50 75 100%
Actual concentration
of ions on resin at
corresponding height

0 25 50 75 100%
Actual concentration
of ions in water at
corresponding height

Figure 11.4. Actual concentrations of ions within ion exchange zone (schematical)

Regeneration

When the unit is exhausted, it has to be regenerated with strong NaCl brine and the normal practice is to pass this down the column in the same direction as the water flowed during the exhaustion stage. The brine first contacts a part of the bed which is substantially in the Ca^{++} form, and therefore removes Ca^{++} ions which then pass on down the column. When this Ca^{++} rich solution gets to the bottom of the column, there is some tendency for the Ca^{++} to displace Mg^{++} and even more to displace any Na^+ which may have remained behind at the very bottom. A regeneration zone is formed which now travels down the column in exactly the same way as the ion exchange zone. It carries in front of it a Ca^{++} rich solution which tends to expel all the less tightly held ions from the resin, though it must be remembered that the affinity differences have been greatly reduced because of the much higher concentration at which regeneration takes place. However, the fact is that pure NaCl brine only comes into contact with resin after a Ca^{++} rich solution has already gone over it, and has left some Ca^{++} on the resin. One important difference between the exhaustion and regeneration stages is that in exhaustion the equilibrium is reached very quickly, while the regeneration equilibrium takes some time. In regeneration, therefore the zones are much less sharply defined.

Conclusions

What are the conclusions to be drawn from this description of what goes on inside an ion exchange softener?

First, the bed of the exchanger always remains predominantly in the Ca^{++} form. It is delivered by the manufacturer in the Na^+ form, but after the first exhaustion it never returns to the all-Na^+ form again. As we saw in Figure 11.2 only an enormous excess of NaCl brine would be capable of achieving total regeneration. In practice, only about one-third of the resin's total capacity is used; the rest is permanently in the Ca^{++} form.

Second, a resin which has most of its groups in the Ca^{++} form even when it is regenerated will come to equilibrium with water which still contains a trace of Ca^{++}. Theoretically it is possible to obtain a totally soft water only by contacting it with a resin in the all-Na^+ form. In practice, however, the leakage of Ca^{++} is merely held to an acceptable level.

Third, the practice of regenerating in the same direction as exhaustion drives Ca^{++} down the bed and puts Ca^{++} ions on to the bottom of the bed, thus increasing the leakage in the next cycle.

This last point has been obvious to many workers in ion exchange for decades. However, downward regeneration is so much simpler than upward regeneration that it remains the common method and it is capable of yielding an acceptable quality of water at a reasonable cost. In recent years, however, good progress has been made in regenerating the bed upwards, that is in counterflow. We shall look at the advantages and disadvantages of this in Chapters 14–18.

12 Dealkalization and weakly acidic resins

In Chapter 9 the bicarbonate ion HCO_3^-, which is called alkalinity in water analyses, was described as the joker in the pack. In practice, alkalinity is no joke at all and causes serious trouble in boilers. If there is much of it in the feed water, it must be taken out, even for low-pressure boilers.

The HCO_3^- ion exists as a link in a chain of reversible reactions which start with CO_2 in the atmosphere dissolving in water as shown in the reaction chart in Figure 12.1.

$$\begin{array}{ccccccc}
& \text{Reaction 1} & & \text{Reaction 2A} & \text{Reaction 2B} & & \text{Reaction 3} \\
CO_2 & \xrightleftharpoons{\qquad} & CO_2 & \xrightleftharpoons{\qquad} H_2CO_3 & \xrightleftharpoons{\qquad} H^+ + HCO_3^- & \xrightleftharpoons{\qquad} 2H^+ + CO_3^{--} \\
\text{atmosphere} & & \text{in water} & & &
\end{array}$$

Figure 12.1. Reactions of CO_2 and water

Reaction 1 is a case of gas dissolving in water according to Henry's Law (see p.137). This says that the amount of gas dissolved is proportional to the partial pressure of the gas in the atmosphere above the water. Rising temperature changes the Henry's Law constant and reduces solubility.

Reactions 2A and 2B can be lumped together for practical purposes. Carbonic acid H_2CO_3 has such a small solubility that it plays no part in its own right. For simplicity we can take the two together to read:

Reaction 2

CO_2 in water $\rightleftharpoons H^+ + HCO_3^-$

H_2CO_3 is a weak acid and not always totally dissociated (see Chapter 10). Unlike a strong acid such as HCl, the fraction which dissociates depends on the availability of all the components in the reaction; of these, the H^+ concentration is the factor most likely to vary. At high pH, when the H^+ concentration is low, the reaction goes well over to the right, whereas in acid solution it hardly goes at all, and all the CO_2 stays as a dissolved gas.

Reaction 3 starts with HCO_3^- ions and unless they are present it cannot of course go forward at all. In fact, it needs a very alkaline solution (which will be greedy for H^+ ions) to give rise to any quantity of CO_3^{--} ions.

Each of these three reactions can go both forwards and backwards and therefore ends up in a state of equilibrium, which depends on the availability of all the components which take part in the reaction. A change in the availability of any of the components will cause a change in the equilibrium of one reaction and this sets up changes in all the other reactions, right through the whole chain.

The rules which govern these equilibria are quite complicated but fortunately engineers and plant chemists need not fuss with them at all. A simplified convention has developed from the use of methyl orange and phenolphthalein indicators, which are used in analysis and change colour at pH3.8 and 8.3 respectively. For normal use the convention says that below pH3.8, the HCO_3^- ion does not exist at all because Reaction 2 does not go forwards, and below pH8.3 the CO_3^{--} ion does not exist at all because Reaction 3 does not go forwards. Between pH3.8 and 8.3 the relationship between dissolved CO_2 and the HCO_3^- ion depends on the pH and is shown in Figure 12.2 whose graph only applies to cold water in which the hardness is greater than the alkalinity, which is normal in natural waters.

The convention serves well for industrial use but involves minor limitations and inaccuracies.

Getting rid of the alkalinity (dealkalization)

The chain of reactions which is involved starts with CO_2 gas at one end and finishes with the CO_3^{--} ion at the other; both these are easily removed from solution. CO_2 is sparingly soluble in water and can be blown out if the water is brought into contact with an atmosphere in which the concentration of CO_2 is low. This can be done in an air-blown degasser tower or in a steam-scrubbed heater or deaerator. At the other end of the equilibria, the CO_3^{--} ion forms insoluble compounds with Ca^{++} ions, which can be precipitated in controlled conditions in a lime softener.

Any interference in the equilibria which will make them move entirely to the right or to the left can therefore be used to remove alkalinity.

The oldest external water treatment process was to make the feed water boil

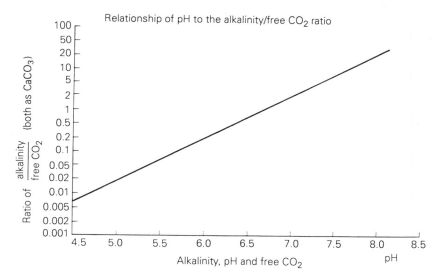

Note: The pH of a water is determined by the ratio of alkalinity to free CO_2. The graph applies at ambient temperatures to waters which do not contain sodium bicarbonate. If the alkalinity exceeds the total hardness, then sodium bicarbonate is present and this graph does not apply.

Figure 12.2. Alkalinity, pH and free CO_2

in a tank by blowing steam through it. The rise in temperature reduces the solubility of CO_2 which therefore comes out of solution and pulls Reaction 1 to the left. When Reaction 2 goes to the left in consequence, H^+ ions are taken out of solution which makes the pH rise, so that Reaction 3 goes to the right. The CO_3^{--} ion which is produced precipitates out with the temporary hardness. These are the reactions which we want to prevent from taking place in the boiler; the Victorians deliberately promoted them in what was called a de-tartarizer.

Lowering the pH

In this section we are concerned with ways of moving all the reactions to the left by lowering the pH (see Figure 12.3).

The most obvious way of doing this is to dose with a strong acid such as H_2SO_4. Suppose we have a water of pH7.4 with an alkalinity of 229ppm. According to Figure 12.2 this water has about 45ppm of free CO_2 in it. Acid introduces additional H^+ ions, which push all the reactions to the left. The pH falls, but Reaction 2 takes H^+ ions out of solution as it goes to the left, so

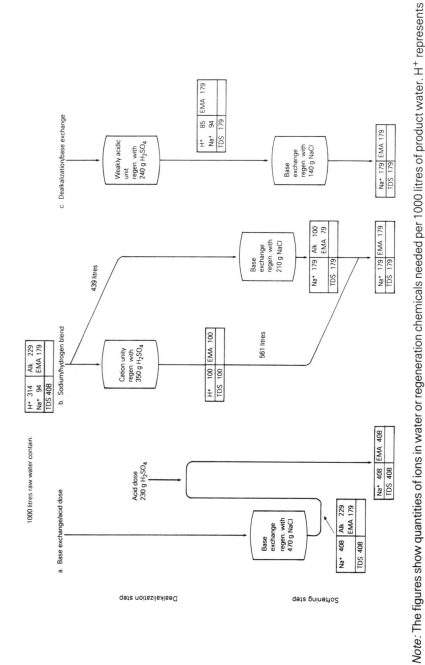

Note: The figures show quantities of ions in water or regeneration chemicals needed per 1000 litres of product water. H^+ represents hydrogen ions. H represents hardness. All quantities are expressed in grams equivalent of $CaCO_3$.

Figure 12.3. Three methods of obtaining a soft water of zero alkalinity

that the fall is much smaller than would be the case if acid were dosed into pure water; this effect is called buffering. Figure 12.2 shows that a dose of 100ppm brings the alkalinity down by an equivalent amount, but the pH only falls to about 6.5 (Alk/CO_2 ratio 129/145), while such a dose in an unbuffered water would have produced a pH of about 2.7: 224ppm of acid reduces the alkalinity to 5ppm but the pH is still about 5.0. With 229ppm of acid we reach zero alkalinity at pH3.8.

If the CO_2 is then blown out of this water, we have removed the buffering. pH3.8 in an unbuffered water represents 8ppm as $CaCO_3$ of free acid so that a very small dose of caustic soda will neutralize this and bring the water back to pH7.

Acid dosing is quite a good way of removing alkalinity but there are ways of achieving better results.

Suppose that the complete analysis of the water which we are treating (a typical London water) is:

Hardness (Ca^{++} and Mg^{++})		314ppm as $CaCO_3$
Na^+ and K^+		94ppm as $CaCO_3$
	Total	408
Alkalinity		229ppm as $CaCO_3$
SO_4^{--} NO_3^- and Cl^-		179ppm as $CaCO_3$
	Total	408

Instead of adding additional H^+ ions to this water by acid dosing, we can get the same result by exchanging some of the cations in it for H^+ ions. Suppose the water passes over a cation exchanger in the H^+ form and the exchange is controlled so as to swap just 229ppm of the cations for H^+ ions. The effect on the alkalinity is the same as adding such an amount of acid, but instead of putting more stuff into the water we have greatly reduced its total dissolved solids, as shown in Figure 12.3.

Ion exchange techniques

We now face the problem of controlling the exchange of cations for H^+ in such a way that it just equals the alkalinity in the water. Any less than this will leave alkalinity behind because it will fail to convert all the HCO_3^- ions to CO_2. Excess H^+ ions, which find no HCO_3^- ions to combine with, remain behind as free mineral acidity (FMA). Not only does this represent a waste of acid but FMA uses up extra caustic soda when the treated water is neutralized to make it non-corrosive. The ideal process leaves a few ppm of alkalinity behind.

A cation resin has sulphonic groups giving it ion exchange properties and these are as strong as sulphuric acid, which they resemble chemically. If water is passed over normal (strongly acidic) cation resin in the H^+ form, then all but a trace of the cations in the water will be exchanged for H^+.

But although such a resin knows no moderation, the plant designer can use it in moderation. If, in the example above, out of every 1000 litres of our London water he passes only 561 litres through the cation exchanger and then blends the acid product with the other 439 litres which have bypassed the cation unit, the quantity of H^+ ions introduced into the blend just equals the alkalinity. The H^+ combines with this alkalinity to form CO_2 which can be eliminated from the solution, and all that is left after blending is a quantity of total dissolved solids (TDS) equivalent to the original EMA.

This process was much used in the USA, combined with base exchange, as the sodium/hydrogen blend process, which is shown in Figure 12.3.

Europe, on the other hand, has led in the progressive development of resins which use the weak acid group $-COOH$, called carboxylic acid. For dealkalization, these weakly acidic resins have three great advantages over sulphonic resins:

- the active group is an acid of about the same strength as carbonic acid;
- these resins can be made with a higher capacity than sulphonic resins and with a greater proportion of the total capacity available for use;
- the weakly acidic resin does not hang on to cations so tenaciously and therefore its regeneration needs a much smaller excess of acid. In practice, the ratio of acid used:ions exchanged need not exceed 1.05 compared with sulphonic resins which usually need 1.5 or more.

The use of weakly acidic resins

The acidity of the carboxylic group $-COOH$ varies slightly in different resins but for simplicity we can assume that it is exactly as strong as carbonic acid, which it resembles chemically. In terms of our convention this means that RCOOH only dissociates to $RCOO^-$ and H^+ when the pH of the water in which the resin lies is above 3.8 (in this formula, R stands for the resin matrix, which is an inert lattice).

Above pH3.8, with its active groups dissociated, the resin performs generally like a strongly acidic resin, giving up its H^+ in exchange for cations and showing a particular preference for the divalent ions Ca^{++} and Mg^{++}. When London water passes over a bed of this resin, cation exchange will take place until the amount of H^+ ions released into the water just equals the original alkalinity. At this point the pH reaches 3.8 and the resin goes to the inert form of RCOOH. According to this simple view, therefore, a carboxylic resin automatically removes the alkalinity from the water without creating

FMA. Because of its preference for divalent ions this cation exchange is for hardness, with the result that the temporary hardness is removed at the same time, to give the quality of water shown in Figure 12.3. This is a better quality of water than can be obtained by simple hydrogen blend because more of the hardness is removed.

In practice, the resin does not work with quite such precision. Over-regeneration with acid can make it give a product of up to 20ppm FMA; under-regeneration will result in substantial slip of alkalinity. (In a way it is misleading to speak of a slip of alkalinity, since the resin does not react directly with the alkalinity at all. To be exact we should say that the resin fails to exchange enough cations to equal the alkalinity in the water.) However, these extreme conditions can be avoided by automatic pH control of the regeneration cycle.

Dealkalization and base exchange

The removal of temporary hardness and blowing out the CO_2 formed (see Chapter 15), is usually not enough. Once it is worth putting in an ion exchange plant, it is usually worth removing all the hardness from the water.

Where alkalinity is removed by acid dosing, this means conventional base exchange, usually before acid dosing, to avoid corrosion on the exchanger apparatus, as shown in Figure 12.3.

With hydrogen blend (Figure 12.3), all that is necessary is to instal a base exchanger in the line which bypasses the cation unit. The acid water from the one stream contains free H^+ ions which combine with the HCO_3^- ions which remain in the softened water of the other stream and thus eliminate themselves from the solution. All that is left after blending is a TDS equivalent to the original EMA, and the water is totally soft.

Dealkalization by weakly acidic exchange achieves the same reduction in TDS as hydrogen blend but leaves the permanent hardness behind. To achieve a totally soft water the product of the weakly acid exchanger must be passed through a base exchanger, as shown in Figure 12.3. Although this unit must be sized to take the full flow of water, the amount of hardness which it has to remove is actually less than the amount removed by the base exchanger in the sodium/hydrogen blend process.

These examples show the actual quantities involved in treating 1000 litres of London water by the three processes and they can be summarized in Table 12.1.

Table 12.1.
Comparison of treatment systems for softening a typical water

	Base exchange/ acid dose	Sodium/hydrogen blend	Dealkalization/ base exchange
Acid used*	230g	350g	240g
Salt used*	470g	210g	140g
Running cost	High	Middling	Low
Capital cost	Low	Fairly high	High
Treated water quality	High TDS	Good	Good
Auto. control	Simple	Difficult	Simple
Effluent	Neutral	Very acid	Slightly acid

*The figures refer to the example shown in Fig. 12.3 but are fairly typical. They represent grams as $CaCO_3$ per 1000 litres treated water.

13   The principles of demineralization

Every engineer knows that the higher the pressure and temperature of a boiler, the greater the need for pure boiler water. What is more, the modern tendency for high heat transfer rates, which are being adopted in order to reduce the cost of the boiler itself, means that still purer waters have to be used. This is because the intensively used heat transfer surfaces lose efficiency much more rapidly with slight scale formation, and also because the tubes are vulnerable to the resultant overheating.

Many textbooks give guides for boiler water quality, such as for example BS2486 of 1954 (about to be replaced). In water treatment, however, we are more directly concerned with the feed water and with the make-up water, which generally have to be much purer than the boiler water. Figure 13.1 shows a general idea of permissible boiler water contents. From these a mass balance can be calculated, using the proposed proportions of feed, blow-down, recycled condensate (if any) and make-up.

Very often the make-up has to be so pure that it can only be obtained by ion exchange demineralization. Actually demineralization is a misnomer; the ion exchange process can never remove anything but must always be an exchange. In demineralization, the salts removed from the water are exchanged for more water.

The chemical principle

In the pecking order of affinities which we discussed earlier (see p.93), the ion held least tightly by a cation exchanger is the H^+ ion. If a raw water

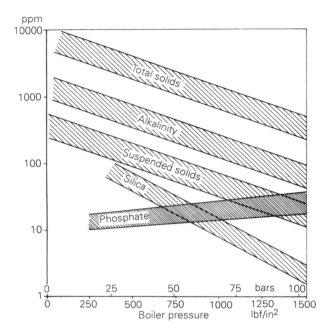

ppm

Note: The quantities indicated in this graph are only intended as a very rough guide, as actual values depend greatly on the boiler as well as the boiler pressure.

Figure 13.1. Maximum permissible boiler drum contents

is passed over a cation exchanger in the H^+ form, all the cations in the water will tend to exchange with the H^+ ions on the resin. The water leaving the ion exchanger will then contain practically no cations except H^+ ions; the anions will of course have remained unchanged, and the water is therefore markedly acid.

If this acid water is now passed over an anion exchanger in the OH^- form, the OH^- ion too is the least tightly held of all the anions concerned so that the anions from the water will tend to go on to the resin in exchange for OH^- ions.

We have explained that H^+ and OH^- ions are subject to a rule which limits the concentration in which they can exist together. As soon as the OH^- ions from the anion exchanger get into the water, this permitted concentration is exceeded and the OH^- ions therefore pair off with the H^+ ions to form water.

$$H^+ + OH^- = H_2O.$$

In the end, if all the salts are taken up by the ion exchanger, all the H^+ and

108

OH^- ions from the resin end up as water. The salts have therefore been exchanged for an equivalent amount of water and the water demineralized.

The problem of leakage

The overall demineralization reaction can be achieved by performing one ion exchange reaction after the other or by carrying on both simultaneously. The problem is to overcome the equilibria which cause leaking of ions.

Ion exchange takes place because the resin has a higher affinity for some ions than for others. This preference is never absolute; it is a matter of degree. Suppose we take an NaCl solution of, say, 300ppm as $CaCO_3$, and pass it over a cation exchanger which is partially in the H^+ form. The resin prefers Na^+ ions and will therefore exchange them for H^+ ions, but the total concentration of cations in the water will remain the same — for every Na^+ ion taken out of solution, an H^+ ion is put back into it. These H^+ ions attempt to get back on to the resin, and succeed in increasing numbers as the concentration of H^+ ions in the water increases. Finally we reach an equilibrium in which there are so many H^+ ions in the solution trying (rather feebly) to get back on the resin in competition with a very small number of residual Na^+ ions that the return of H^+ on to the resin cancels out the uptake of Na^+. As in all equilibria, chemical reaction only seems to have stopped but in fact is going backwards at the same speed as it is going forwards.

The two reactions which make this equilibrium can be written in the form:

(1) H^+ on resin + Na^+ in water $\rightarrow$ Na^+ on resin + H^+ in water.
(2) Na^+ on resin + H^+ in water $\rightarrow$ H^+ on resin + Na^+ in water.

Equation 1 represents the reaction which is doing useful work and Equation 2 is the one which is defeating the object. As long as reaction 2 proceeds at all, there must be some Na^+ left in the water, but the slower it proceeds, the smaller will be the remaining slip of Na^+ in the product.

The way to slow down or stop this second reaction is to reduce or eliminate the availability of one of the two reagents on the left-hand side of the equation. If only we could achieve a resin containing no Na^+ at all, or a water with no H^+ in it, we could theoretically obtain a product with no residual Na^+ at all. One way of achieving a high purity of product, therefore, is to convert the resin completely to the H^+ form at the point at which the product water leaves the apparatus, which is where the final equilibrium is reached or at least approached. But even at the high concentrations at which it is regenerated, the resin has a higher affinity for Na^+ than for H^+, so that unless we are rather clever this would require regeneration with a huge and uneconomical excess of acid.

109

Plate 9. Small packaged demineralizer comprising cation and anion vessels capable of producing quite high-quality water

In the treatment of natural waters, the situation is even more difficult because natural water supplies contain mostly Ca^{++} and Mg^{++} which are even more tightly held by the resin than Na^+. Whereas the total regeneration of a resin from the Na^+ form to the H^+ form is difficult but possible, the total regeneration of a resin from the Ca^{++} form to the H^+ form is practically impossible. This means that we only have much hope of getting a resin substantially regenerated to the H^+ form if Ca^{++} and Mg^{++} are kept from it in the exhaustion stage. Counterflow regeneration is the clever way to achieve this condition but it raises difficulties of its own (see p.119).

110

Plate 10. Typical demineralization plant in a brewery

Two important consequences flow from these facts: as Na^+ is the least tightly held of the cations found in natural waters, the slip from a cation unit is always of Na^+ rather than Ca^{++} or Mg^{++}; but this slip can be reduced, all other things being equal, by increasing the amount of acid used to regenerate the resin.

111

Plate 11.   Major ion exchange plant at a nuclear power-station

Attacking both ions together

The other reagent necessary to further the unwanted reaction above is H^+ in water; if this could be removed before cation exchange is complete, then the slip of Na^+ would be very small. But this is just what happens in the OH^- form anion exchange unit, as we saw earlier in this chapter; the anions in the water are exchanged for OH^- ions, and these immediately pair off with excess H^+ ions, forming water and removing themselves from the fray.

It is thus possible to obtain very high quality by passing the water through the cation exchanger in two stages, with an anion exchanger between them and a second anion exchanger at the end. Figure 13.2 shows the flow sheet of such a process. For simplicity it is assumed to be treating the water of 300ppm NaCl as $CaCO_3$ which we considered earlier.

The first cation unit removes, say, 280ppm Na^+, leaving 20ppm Na^+ and of course all the 300ppm Cl^- to pass on. The anion unit exchanges the Cl^- for OH^-; the first 280ppm of Cl^- exchange yields OH^- ions which immediately pair off with the H^+ ions already in the water, and only the

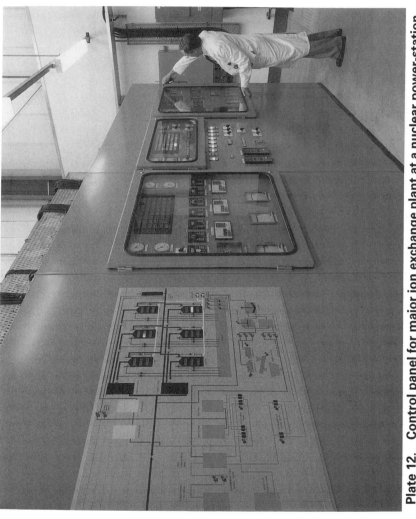

Plate 12. Control panel for major ion exchange plant at a nuclear power-station

Plate 13. Typical demineralization plant at a major power-station

exchange of the last 20ppm of Cl^- ions effectively leaves corresponding OH^- ions in the water. With such a small concentration of OH^- ions trying to get back on to the resin, the slip of Cl^- is correspondingly small. In fact the water leaving this anion unit would probably contain 19ppm OH^- and 1ppm Cl^-. Bearing in mind the 20ppm Na^+ still in this water, we now have a very dilute solution of caustic soda, plus a trace of salt.

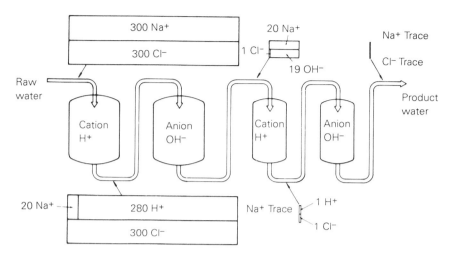

Note: The concentration of ions at various stages of the process is illustrated by areas which demonstrate their equivalent concentration.

Figure 13.2.   Classical demineralization flow sheet

In practice of course natural waters are not all NaCl but, even if we had started with a natural water, Na^+ would be the ion which leaked through the cation unit, and therefore the water coming out of two units of this kind would again be mostly Na^+ and OH^-.

If this water is passed on to the second cation exchanger, once again the first 19ppm Na^+ exchanged result in H^+ ions which pair off with the OH^- ions already there and disappear from the scene, so that the reverse reaction only appears as a result of the resin exchanging H^+ ions with the last ppm of Na^+. This naturally leaves no more than a trace of Na^+ behind. In practice this operation is quite easy to perform, because when treating water the second cation only receives the leakage, which consists of Na^+ ions, and these are relatively easily regenerated off to give a resin with a high regeneration level.

The water from the second cation unit still has 1ppm Cl^- which is then almost completely exchanged by the second anion unit, yielding a very high quality demineralized water.

Unless a totally demineralized water is fed into an ion exchange unit (which of course rarely happens in practice), the water coming from a cation exchanger is always going to contain an excess of H^+ ions and will therefore be acid, while the water leaving an anion exchange unit is always alkaline, as long as the units are not exhausted. For this reason it is impossible to treat natural waters by exchanging anions first because the high pH this would generate would precipitate the Ca^{++} and Mg^{++} in the water.

115

Mixed beds

The flow sheet cation-anion-cation-anion, which is shown in Figure 13.2 is one which used to be much favoured in Europe and has given very good service in the past. Its main disadvantage is a relatively high capital cost.

If the water could be passed through more cation–anion pairs then the chemistry would be the same but the purity to which the water could be demineralized would be even greater. A mixed bed of cation and anion resins effectively does this; it is the best way of getting ultra-pure water and is commonly used for that purpose. However, mixed-bed operation involves some mechanical difficulties and its process chemistry is not quite as straightforward as it seems at first glance. Nevertheless, the flow sheet cation–anion–mixed bed is standard and costs substantially less than the arrangement shown. Chapter 17 will deal with the problem of mixed beds.

Either way, by the use of repeated cation–anion exchange, it is possible to obtain water of very high quality. The fact that such very high purities are now obtainable has made it possible for boiler makers or users to specify maximum feed water contents so small that it is almost impossible to visualize them. For example, a sub-critical 500MW power-generating set contains about 500 tonnes of water in circulation round the boiler and turbine. This feed water is required to contain less than 0.01ppm (or 10ppb) of the main impurities such as iron. This mass of water would fill a tank 5m diameter by 25m long, and 10ppb of iron in it is about the weight of a 2BA nut. Higher pressure and nuclear boilers have even more stringent specifications.

Some chemists specify water purities without really considering what they are asking for and why they really need such very pure water. Fantastic quality of this kind is not to be had without spending money and taking pains.

Part 6
ION EXCHANGE UNITS AND SYSTEMS

14   Counterflow

The previous chapter has we hope created an understanding of the chemistry of ion exchange. In the practical world we have to put this to work for us in an economical manner.

An ion exchange unit is essentially a package designed to contain ion exchange resin and allow intimate contact between this and fluid to be treated. It also has to allow intimate contact between the resin and whichever regenerant is used to treat it when it is exhausted.

The unit is normally a pressure vessel made of steel or glass-reinforced plastic suitably designed to withstand the operating pressure. It can be designed to a code of construction such as BS1500, ASME or DIN and its internal surface must be protected against corrosion by lining it, frequently with rubber, if made of steel.

The particular skills brought to the situation by the designer relate more to the design of distributors and specific areas of expertise such as ion exchange kinetics than the general engineering parameters also necessary.

The easiest form of ion exchange unit to design is one where the water flows from top to bottom and, when the resin is exhausted, the regenerant does the same thing. The designer has to cater for very different flow rates between the water and regenerant and ensure that good distribution is achieved with each. Sometimes he has separate distributors for each of the flows if he cannot match the turndown ability of his distributor to the differing requirements. He may, on the other hand, increase the flow rate of the regenerant by diluting it beyond optimum strength.

In Figure 14.1 during service, valves 1, 3 and 6 would be open. During

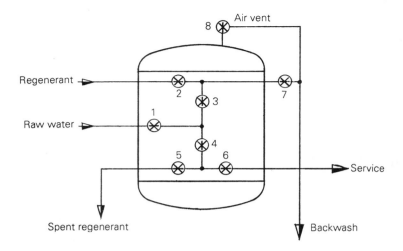

Figure 14.1. Typical unit (co-flow)

regeneration, valves would be opened as follows: backwash valves 1, 4 and 7; regenerant injection and slow rinse valves 2 and 5; fast rinse valves 1, 3 and 5. The air vent would be used probably once a cycle to get rid of accumulated air.

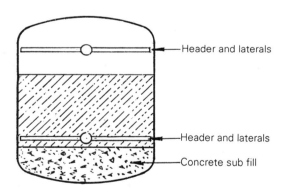

Figure 14.2. Typical unit internals

Internally the unit could look like Figure 14.2.

In a simple unit the sub-fill could be of graded gravel and the exit header could be replaced by a simple cone in small units.

A more sophisticated bottom distributor would normally comprise a nozzle plate with strainer nozzles fitted as shown in Figure 14.3.

120

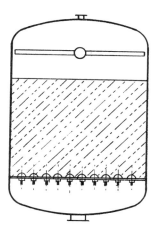

Figure 14.3.   Nozzle plate unit design

Such units are very adequate where minimization of running costs and optimisation of quality are not of prime importance.

These co-current types have the disadvantage that the strong regenerant drives the contaminants (Ca, Mg, Cl, SO_4, depending upon the type of unit) before it and leaves the bottom of the bed contaminated. This problem would be avoided if the regenerant started at the opposite end. Then the strongest regenerant would treat the least contaminated part of the resin first, leading to better quality on the next run.

Counterflow regeneration (CFR)

In the ion exchange units described the water being treated flows downwards through the resin bed, and the bed is then regenerated in the same direction. This is a practice which not only lowers the efficiency with which the regenerant is used but increases the unwanted leakage of ions in the treated water.

The remedy has long been obvious: to regenerate in counterflow, that is to treat water in downflow through the bed and then regenerate in upflow, or vice versa. But though the remedy was obvious, the actual means of achieving good counterflow regeneration have raised such problems that commercially acceptable solutions took some years to evolve. ('Countercurrent' has been loosely used to described this method of operation but this is an incorrect use of the word. In chemical engineering practice, 'countercurrent' means simultaneous flow in opposite directions such as, for example, is found in degasser towers, in which water flows down while air is blown up through the tower at the same time.)

121

Mechanical problems

Good contact between liquid and resin is only obtained when water flows through a packed and immobile bed. One of the objects of CFR is to put to good use the chromatographic bands in which ions end up on the resin: mixing is therefore particularly to be avoided. In CFR, however, either service or regeneration must take place in upflow, which normally lifts the bed, increases the void space and causes the resin to fluidize, circulate and mix. The most immediate problem therefore is how to keep the ion exchange bed packed and immobile during upflow.

One obvious device would be to fill the unit brim-full of resin so that the bed remains packed all the time. This is impractical except on very small units, because the resin swells and shrinks during the operational cycle and will either damage itself if it has no space in which to swell or, when the resin shrinks, it will create a space which allows the bed to move and lose its orderly banding. Another problem is that dirt and broken resin particles should be backwashed out from time to time, which can only be done from a bed which is free to move.

Some designers have used upward service and downward regeneration and have met with some success. Most CFR designs in the UK, however, use downward service flow for units over about 1m dia.

A successful design has to meet all the normal requirements of conventional ion exchange units, plus some additional ones. The requirements therefore include the following:

- the unit must have the normal arrangements for water to flow downwards through the bed during service;
- there must be provision for backwashing from time to time to re-grade the bed and remove dirt and fines. This calls for the bed to be free to expand by 50–100 per cent of its settled volume;
- the resin must be free to shrink and swell during the cycle (normal volume changes are less than 10 per cent, but in special applications they may be a good deal more);
- there must be a means for holding the bed packed and immobile after service and during upflow regeneration;
- like all other water treatment plant, the apparatus must be cheap, simple, robust and corrosion resistant.

Solution to the mechanical problem

It is fairly easy to meet these requirements in small units of less than, say, 1m diameter. On the normal industrial scale, however, things get much more difficult. Many different devices have been suggested for CFR operation but very few of these have proved successful on the large scale.

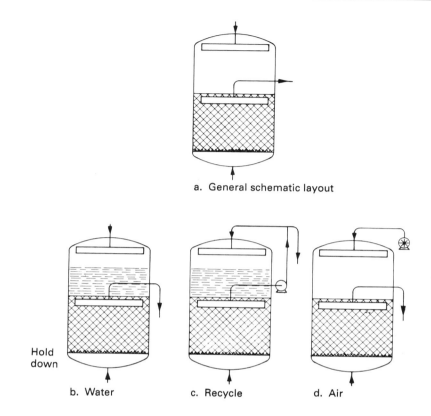

a. General schematic layout

Hold
down

b. Water   c. Recycle   d. Air

Figure 14.4. Typical liquid and air hold down designs

One group of designs which relies on the same basic principle has been successful in the field are as shown in Figure 14.4. The ion exchange unit is of conventional layout and consists of a resin bed in a cylindrical vessel above which there is the usual 50–100 per cent rising space. The top and bottom of the unit are fitted with inlet and outlet distributors. In addition to these, however, there is a third, a buried collector lying just below the upper level of the resin bed and fitted with strainer nozzles or mesh-covered holes through which resin cannot pass.

These units are run in normal downflow service. When the resin is exhausted, regenerant is pumped into the bottom distributor and passes upwards through the bed. At the same time, air or water enters the top distributor and flows through the very small depth of the bed which lies above the buried collector. Both the two flows leave the unit through this buried collector. The upflow of regenerant tends to make the resin lift and mix but

123

this tendency is overcome by the force exerted on the bed by the downflow through the upper layer.

This basic principle has been used in several variations. The liquid which enters the top distributor during regeneration may be water (Figure 14.4b) but this has the disadvantage that it uses a large quantity of water and creates a large volume of dilute waste. The downflow may also be of spent regenerant (Figure 14.4c) which avoids the waste of water but needs an expensive corrosion-resistant pump. The most successful variation uses air (Figure 14.4d) which avoids the dilution problem. If this device is used in an ion exchange train at the end of which there is a mixed bed (see Chapter 17) then the train must in any case include an air blower for air mixing, and this blower can double up for the CFR duty at no extra cost.

Air is very efficient as a medium for holding the bed down, because its use means that the layer of resin above the buried collector is in a semi-dry condition and its weight lying on the rest of the bed adds an important component to the forces holding the bed down. The use of air for this purpose was first tried in Europe many years ago and has been widely successful; the practice is beginning to be adopted in the USA.

The resin above the buried collector is never regenerated and plays no part in the ion exchange process; on the contrary if this totally exhausted resin should accidentally get into the working portion of the bed it could have a damaging effect on CFR operation. For this reason, and for economy, a layer of inert resin which is lighter than ion exchange resin is sometimes used to constitute the uppermost layer of the bed (Figure 14.5a). The buried collector may be of the header-and-lateral type (Figure 14.5b) or the collector may be placed outside the bed with long individual downcomers dipping into the bed, each downcomer being fitted with a strainer nozzle (Figure 14.5c). The last design avoids the heavy mechanical strain which is put on a buried header and lateral when it is placed within the resin bed.

Any suspended matter contained in the water tends to filter out at the top of the bed, and an upward wash from the buried distributor will remove it. This upward wash does not disturb the rest of the bed and the chromatographic banding within it, and generally speaking makes regular backwashing of the whole bed unnecessary.

Hydraulic problems

CFR presents two sets of hydraulic problems. The first is the design of the unit in such a way that the bed is reliably held down during regeneration. The various component forces exerted by the upflow and downflow have proved impossible to measure and the physical layout of the buried collector also plays a part in holding the bed. Successful design of CFR units therefore depends on experience.

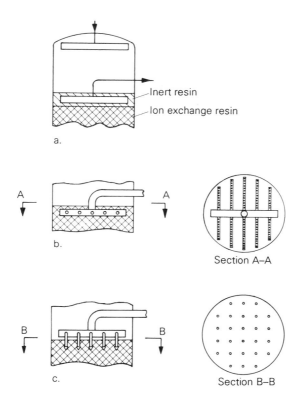

Figure 14.5. Typical regenerant collector systems

The other problem is that of liquid distribution, which is much more acute than it is in the design of conventional units. The object of CFR is to use regenerants more efficiently, and therefore in the nature of the process the amount of slack available is less and the effect of minor maloperation much more noticeable.

In conventional downflow regeneration, the regenerant enters the unit through a distributor and then passes through free liquid space before reaching the top of the ion exchange bed. Any tendency for maldistribution of the regenerant in the top distributor therefore has a chance to correct itself before the chemical enters the resin bed. In upward regeneration, on the other hand, the regenerant is injected directly into the bed and therefore the tendency to redistribute the regenerant is very small.

In both these cases the regenerant enters the unit through a distributor which also has to handle the normal service flow, which will be at a much higher rate.

The distributor therefore has to be designed to handle two flows one of which may, in extreme cases, be as much as fifty times as great as the other. This is an old difficulty in ion exchange design but it is only in CFR design that it has really become acute.

The solution depends on the actual quantities involved in the application. If the two flows are not too greatly different from one another, very careful design of the regenerant-inlet/water-outlet will overcome the problem. In extreme cases, however, some designers use ingenious distributors which have a higher pressure loss characteristic in regenerant upflow than they do in service downflow. This means that they present sufficient resistance in regenerant injection to give good distribution, without causing an excessive pressure loss when used to convey a much larger flow during service. Another solution is to install two separate distribution systems at the bottom of the unit, one for each duty. Each of these two devices raises extra cost.

All distribution problems are eased if the volume of resin needed is put into a relatively tall, narrow bed. Such a bed has a smaller cross-section over which the liquids have to be distributed, it presents a higher pressure loss and therefore assists redistribution within the bed, and the greater linear height extends the chromatographic bands and gives them a better chance to develop. CFR units should always be designed as tall and narrow as circumstances permit.

Chemical problems

Greater chemical efficiency means that less surplus regenerant is being used, and therefore less excess regenerant will remain in the spent solution leaving the unit. This leads to chemical problems of which the commonest arises in the sulphuric acid regeneration of cation units.

Normal waters contain more Ca^{++} than any other cation, so that with H_2SO_4 regeneration the spent regenerant after cation exchange will contain a large proportion of calcium sulphate:

$$R\ Ca^{++} + 2H^+ + SO_4^{--} \longrightarrow R(H^+)_2 + Ca^{++} + SO_4^{--}$$

where R represents the ion exchange resin

The SO_4^{--} ion takes no direct part in the regeneration itself, but when sulphuric acid is used it is of course always present. The problem is that $CaSO_4$ is not very soluble and in a neutral solution if the combined concentrations of the Ca^{++} and SO_4^{--} ions exceed about 2000ppm, the ions will combine to form a crystalline precipitate.

Such a precipitate causes the bed or the distributors to clog. Sometimes the precipitation is only very slight and the effect on pressure loss is not noticed, but the crystals redissolve slowly in the next service run and cause extra leakage of calcium. Calcium sulphate precipitation in the plant must therefore be avoided altogether.

There are three ways of avoiding it. The most obvious is to inject the acid at a very low concentration, but this makes for highly inefficient regeneration because the equilibrium is then unfavourable to regeneration. Another way is to inject the acid very quickly, because $CaSO_4$ will stay in supersaturated solution for a short time, and it is possible to hold off its precipitation until the solution is safely out of the unit and in a drain line. Within reason this is always done, and it is not rare to see the spent regenerant from a cation unit turning milky white in the effluent drain as the supersaturated calcium sulphate comes out of solution. However, very quick injection may not allow enough time for the regeneration reaction to proceed with the desired efficiency. If the regeneration is upflow in a CFR process, a very high upward flow rate makes it more difficult to hold the bed down.

The third way is to use a large excess of acid, because this excess has the property of solubilizing $CaSO_4$. But this chemical inefficiency of large quantities of waste acid appearing in the spent effluent is just what we are trying to avoid with CFR. The whole situation is a good example of Murphy's Law (also known as the Unreasonableness of Nature) which says that the more we try to achieve something, the more Nature will see to it that we can't.

With sulphuric regeneration, the overall outcome is that chemical efficiency can hardly be improved by CFR unless the water has an unusual chemical composition with a relatively low content of Ca^{++} ions. On the other hand it does make it possible to obtain a very much lower leakage of cations. Another, minor, chemical problem is inevitable with CFR. The object of CFR is to keep unwanted ions from the bottom of the unit so that the resin with which the treated water last comes into contact is fully regenerated. After regenerant injection, the regenerant has to be displaced by a slow rinse. This must be done whether the regeneration is co-current or CFR, and is actually a continuation of the regenerant contact time. In CFR, to avoid ions removed by the spent regenerant being pushed back to the zones from where they came, the slow rinse must be in the same direction as the regenerant injection itself. If, however, raw water is used for this, it will deposit fresh ions just where they are wanted least, which is at the treated water outlet of the bed, and where they would cause high leakage during the following service stage. For low leakage in the product water, therefore, it is necessary to rinse with treated water which contains no ions to spoil the highly regenerated resin at the bottom of the unit. The small extra cost and complication which this involves is offset by the fact that in CFR rinse volumes are always much lower than in conventional practice.

The packed suspended bed (PSB) design for counterflow ion exchange

As briefly mentioned some designers have concentrated their efforts on units which operate with the raw water flowing upwards in counterflow.

In Britain, the commoner CFR design has so far been one in which the water runs in downflow, and the bed is held packed for upflow regeneration. The PSB process, which is a normal design in much of Europe, is also available as an alternative in the UK and is quite commonly used in smaller units.

The basic process

We have outlined the requirements which a successful CFR process has to satisfy. The most important of these is that the ion exchanger should be in packed column form and remain substantially undisturbed over a large number of cycles, so that the end of the column through which the water leaves starts the run in a highly regenerated state; it should never become fully exhausted, and it should receive the fresh unused regenerant to restore it to its fully regenerated state.

Other requirements, we have said, are that the bed should be allowed to swell and shrink as the resin changes volume during the various stages of the service/regeneration cycle and that there should be a facility for occasional loosening of the bed and for the removal of fines and dirt.

The holding of a bed in a packed state for upflow regeneration can be difficult. The logical mind may well ask, therefore, whether it would not be more sensible to regenerate the easy way, in downflow, and then let the unit run upflow for service.

A well-engineered plant does this quite readily. When a settled, compacted bed is put on backwash, it does not break up immediately. What actually happens is that first the whole bed starts to rise as a piston, and a clear water space opens up underneath the column. The resin at the very bottom then begins to fall away, and the bed starts to break up from the bottom. If, however, the unit only has a minimal rising space, then the rising bed immediately comes up against the top distribution system and clusters against the top; a small portion at the bottom will break away and become fluidized but the bulk of the bed will remain undisturbed. Figure 14.6a shows this schematically.

Naming the process

The Germans called this the *Schwebebett* (from *Schweben*, to be suspended, to hover, to float, and *Bett*, a bed). They could have translated this as hover-bed but unhappily they preferred the misleading 'fluidized bed', which is the

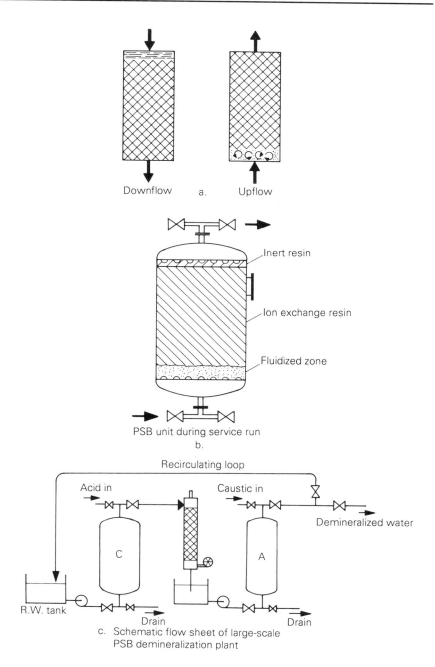

Downflow   a.   Upflow

Inert resin

Ion exchange resin

Fluidized zone

PSB unit during service run
b.

Recirculating loop

Acid in

Caustic in

Demineralized water

C

A

R.W. tank

Drain

Drain

c. Schematic flow sheet of large-scale
PSB demineralization plant

Figure 14.6. The PSB system

proper chemical engineering name for a bed of particles moving freely in suspension. (True fluidized-bed ion exchange designs have been tried, without much success in water treatment, though they show promise for uranium recovery etc.) Realizing their error, they then made things worse by calling it 'packed bed'. Truly packed beds are in wide use on the very small scale where convenience outweighs the problems which can arise if the resin is not given room to swell or shrink. These misnomers can't have helped the process establish itself. We call the design the packed suspended bed, PSB for short.

The PSB unit

To reach its present successful form, the PSB design has had to undergo the usual period of engineering development and Figure 14.6b shows the result.

The main novelty is that the unit has no rising space as we understand it; the top of the settled resin column is only about 100–200mm below the top distributor. The other novelty is that the top distributor is in the form of a nozzle plate, similar to the conventional system at the bottom of the unit. This adds considerably to the cost of an otherwise economically manufactured unit.

Large-diameter units may have a thin layer of special inert resin at the top of the column, made of polyethylene (or some such material) which is inert and floats in water and therefore remains above the ion exchanger. It thus prevents the finest beads from lying against the slots of the top nozzle plate, which raises more pressure loss than the coarse beads lying against the bottom nozzle plate. In smaller units this additional pressure loss is much less important and the complication of an additional resin can be left out.

Preventing the bed from moving

A bed can only become disturbed while it is fluidized in upflow or while it is settling down. In the upflow regeneration designs which have so far been more familiar, things are therefore prone to go wrong during the regeneration stage, which is obviously undesirable.

When the PSB bed is in upflow, it is in service, which is less critical than regeneration. As long as upflow continues, it is held positively against the top nozzle plate. Trouble can only arise either when the bed is first lifted or if it is allowed to drop prematurely.

The first lift needs a minimum flow velocity to ensure that the bed comes smartly up against the top nozzle plate — quite a modest flow rate will do, and it is enough to arrange that when the bed goes on upflow it does so at the normal rated output of the unit. Once the bed is up, a much lower velocity is enough to keep it there. Then, if the flow is interrupted, the bed falls and suffers from disturbance, so that if this happens repeatedly during the service run the product quality will eventually suffer.

Both these difficulties are solved by building the ion exchange train with a recycling loop (Figure 14.6c) which cuts in when the demand drops off. Such a loop has other advantages, which we shall come to later.

At the bottom of the PSB, a fraction of the resin is fluidized during the service run, but this is at the raw-water inlet end, where it does not matter. Here the resin is in contact with the incoming raw water for the whole of the run and, even if the contact between resin and water is somewhat less efficient, the resin and water are bound to be in complete equilibrium by the time the run is at an end.

Unit size, bed depth and pressure loss

If we compare units built to hold a given volume of resin, the PSB unit will be much smaller than either co-flow or upflow regeneration CFR units, both of which also have to contain a large volume of rising space. This means that the shell will cost less, and the units need less headroom, which can be important.

Another feature of upflow service is that it raises a smaller pressure loss than downflow; in upflow the force of gravity and the hydraulic force work against one another and the bed does not become as compacted as in downflow. Moreover, the portion of the bed which is fluidized raises practically no pressure loss at all.

These two factors together mean that in many cases, the PSB unit will have a smaller diameter and a deeper bed than the alternative designs. This makes the shells still cheaper, and in addition it brings process benefits because CFR works best with the deepest bed which can be installed.

Regeneration and rinse

One of the advantages of CFR processes is that they bring the regenerant and rinse straight into contact with the resin column, so that they pass through the bed as a sharp front. In co-flow, by contrast, the regenerant has to find its way through the rising space in which it becomes diluted and diffused. In CFR, as a result, the regenerant actually reaches the exchanger at the concentration at which it is injected, and the volume of displacement and rinse is much reduced.

The PSB design is particularly good in these respects. As it has no rising space, there are no complications of draining down and refilling. The liquid hold-up in the whole unit is near the theoretical minimum, which is the void volume in the resin, that is, half a bed volume.

In practice this allows the use of a very small volume of concentrated regenerant, which is sometimes desirable. For example, HCl can be injected at 10 per cent provided care is taken to minimize the osmotic shock due to this high concentration. At normal regeneration level the volume of acid is then only half a bed volume.

After injection, the dilution water is allowed to continue in the usual manner, in order to displace the regenerant from the bed. In the PSB design this proceeds most efficiently, so after about half a bed volume the concentration of HCl, say, coming out of the bottom of a cation unit starts to fall rapidly.

In about 1.5 bed volumes it is down to a few hundred ppm of HCl. The anion unit behaves in a similar manner but NaOH is harder to rinse down and displacement takes a little more volume.

After displacement, conventional practice would be to rinse the units down to drain in the direction of service flow. If, however, the plant has a recirculating loop the rinse step can sometimes be avoided altogether. If it is possible to extend the displacement stage sufficiently, the whole plant can then be put on recirculation, with both beds going on upflow. The residual Cl from the cation unit is taken up by the anion resin, and the residual Na from the anion unit on the cation resin. Recirculation continues until the treated water specification is reached; in a fully automatic plant the treated water delivery valve opens at this point. The regeneration cycle simply consists of: service; injection; displacement; back to service again. A benefit in water and effluent economy accompanies this simplicity.

Treated water quality and service water source

Treated water must be used for regenerant dilution water in order to get high treated water quality; that is, the cation unit needs decationized water and the anion unit needs demineralized water. The PSB design is so thrifty with water that for convenience both are normally supplied with treated water. The product quality will then be about $1\mu S$ conductivity and 0.01ppm SiO_2.

If no treated water is available for dilution, raw water has to be used for the cation unit. During displacement this water will leave its cations on the resin just where it ought to be very highly regenerated. In the run that follows, therefore, the water leaves the unit in equilibrium with this partially exhausted resin, and the conductivity of the treated water will accordingly be higher.

PSB units need little displacement water, as mentioned above. If only the minimum raw water displacement is used, the actual amount of ions which it puts on the resin will be small and the demineralized product will still be below $5\mu S$ conductivity.

The anion unit cannot be regenerated using raw water, for fear of calcium and magnesium precipitating when it is used to dilute NaOH. Where no treated water is available, the cation unit has to be regenerated first. When it is ready to go on upflow, its output can then be used for regenerating the anion unit, so the regeneration sequence is a little more complicated and takes longer than regeneration using treated water.

Plant size

Either way, the PSB design makes for easy operation, especially when compared with the upflow regeneration designs, which have complex regeneration sequences involving the use of compressed air. The PSB design therefore lends itself to much smaller scale plants than its competitors and it has been incorporated into standard units, the only packaged ion exchange plant capable of giving all the advantages which CFR operation brings.

At the other end of the range there is no size limitation inherent in the PSB design. The biggest units built so far are 3800mm dia (12ft 6in) and treat 820m^3/h (180,000gph) in a single train. Treating water taken from the River Rhine, which is high in TDS and notoriously badly polluted, it yields a water of 1.1μS and 0.01ppm SiO$_2$ after the strong anion resin.

Backwashing

Co-flow units are backwashed after every service run, to loosen the bed and avoid the formation of channels in it. In CFR processes, this is unnecessary because the flow reversal is enough to serve this purpose. On the contrary, backwashing is delayed as long as possible in order to avoid disturbing the bed. Since the only purpose of backwashing CFR units is to remove dirt and resin fines, the secret of successful CFR operation is to ensure that the raw water is clean and that the resin is chosen to give a long life without creating fines. Given these precautions, CFR units run for months between backwashings.

The PSB design does not provide any easy means of backwashing, as the shell has no backwash space. To get the resin clean, some of it will have to come out of the unit unless the plant is so small that the top of the unit can be taken off and the bed scraped. The question of how and to what extent the design should cater for this depends on economic considerations. A large multi-unit plant can be built with a permanently installed backwash vessel common to several PSB units, and the cost of the backwash tanks will be spread over a number of units. On smaller plants, the cost can probably not be justified for this occasional need (the very large plant cited above runs for over a year between backwashings). Many small units will give good service without ever being backwashed throughout the full economic life of the ion exchange resin.

Summary

The PSB design is an alternative means of obtaining the advantages of CFR in ion exchange, with the following additional advantages over the CFR design traditionally used in Britain:

- possible lower capital cost due to smaller units, simpler pipework and valves;
- no need to provide compressed air to hold down the bed;
- simpler operation and automation;
- suitable for small as well as for large units, and available in packaged plant form;
- reliable and consistent high-quality product, approaching mixed-bed quality;
- thriftier in chemicals, waste water and pumping costs.

What can CFR do?

CFR is no panacea, and its application far from universal. Its major benefit is improved product quality and for many applications it is possible to produce water from a CFR cation unit with such a low cation leakage that a second cation stage or polishing mixed bed becomes unnecessary. This results in a small and cheaper ion exchange train, although the individual CFR units are more expensive by virtue of the buried collector and the sophisticated distribution systems which they need. When treating normal raw waters, CFR sulphuric acid regeneration of cation units yields little or no improvement in regenerant efficiency. With high-sodium waters, or when using HCl, the improvement over conventional methods may be significant. All other things being equal, however, a CFR unit always costs more in capital cost than a conventionally designed unit.

In short, CFR is an alternative process with its own advantages and drawbacks. A skilled designer will know when conditions for its use are suitable, and he can often arrange the flow sheet in such a way that the best use can be made of the potential which it offers.

15   Degassing

Degassing is a valuable process used as part of an ion exchange system. As a chemical engineering term it means the removal of a gas from a liquid. In industrial water treatment the word has acquired a special meaning, which is the removal of CO_2 from water by blowing it with air. (Oxygen removal from water, though strictly speaking also a degassing process, is usually called deaeration and this is dealt with in Chapters 26 and 27). We have inserted a chapter on degassing at this stage to make more readily understood the next chapter on flow sheet options.

In ion exchange demineralization, water is treated successively by cation and anion exchange. We have seen that after the cation stage, the water is acid and all the bicarbonate content in it converted to CO_2. In that state a degasser can blow almost all the CO_2 out of solution quite cheaply. If the CO_2 is not removed, however, the rising pH in the anion column re-converts it to bicarbonate, which the anion resin picks up together with Cl^- and SO_4^{--}. Degassing therefore reduces the load on the anion resin, and since degassers are relatively cheap to build and very cheap to run this constitutes a major economy. The degasser stage is only omitted from a demineralization train if the raw water alkalinity is low, the plant is small or if there is some other special reason for it.

After dealkalization by ion exchange, the water is always degassed to complete the process.

The physical chemistry of degassing

Very simple rules govern the physical behaviour of gases. A gas behaves like a room full of bouncing ping-pong balls; the movement of each ball is quite independent of the actual chemical nature of the gas and the rate at which each ball bounces depends only on its temperature. The pressure of the gas is caused by the balls hitting the walls of the room; each ball (or gas molecule) contributes an identical little bit. Suppose, for example, that two gases are mixed in the ratio 2:1 (by numbers of molecules) and suppose the two gases are represented by red and blue balls respectively. As far as their physical behaviour is concerned the two gases are identical; the balls are indistinguishable from one another, except for their colours. Of the total pressure, two-thirds is due to the red balls and one-third due to the blue balls. If we take the blue ones away and leave the red ones behind (everything else remaining unchanged) then the remaining pressure is at two-thirds of the original pressure of the mixture. This pressure of the one gas alone is called the partial pressure of that gas.

In a mixture of gases, therefore, the different gases occupy a common space but each behaves as though the others were not there. Each gas exerts a partial pressure proportional to the number of its molecules in the mixture; the total gas pressure is the sum of partial pressures of all gases present. This total pressure is also dependent on the temperature (which fixes the rate at which the molecules bounce) but, as all gases in a mixture must be at the same temperature, the relative proportions of partial pressures do not change with temperature.

Vapours of liquids, such as water vapour, behave just like gases. At high pressure, however, when the gas molecules are squeezed into a small space and therefore collide more frequently with one another, these rules break down. All water treatment processes are carried out at pressures at or near atmospheric, so that such deviations can be ignored.

Counting the molecules

A litre of cold air contains about 3×10^{22} molecules, and a litre of water contains about 3×10^{25} molecules. Numbers of molecules do not, therefore, lend themselves to what we might call a handy engineering unit.

The common unit in process engineering is to take the weight of a substance (in lb, kg etc) divided by its molecular weight (MW), to yield units called lb mol, kg mol etc. Each lb mol, or each kg mol contains the same number of molecules.

As an example, take 100 000kg of water at 250ppm CO_2, that is, 100 000kg of water containing 25kg CO_2.

$$100\ 000\text{kg of water is } \frac{100\ 000}{18} = 5555\text{kg mols of } H_2O$$

$$25\text{kg of } CO_2 \text{ is } \frac{25}{44} = 0.57\text{kg mols of } CO_2$$

The 'molar' concentration of CO_2 in water is therefore

$$\frac{0.57}{5555} = 102 \times 10^{-6} \text{ mols/mol}$$

that is there are 102 molecules of CO_2 per million molecules of water.

In calculating the molar ratio, the kg cancelled one another out, leaving a simple ratio not dependent on mass, in the same way that ppm is a ratio and does not depend on any system of units. This molar ratio (molar concentration or molarity) is useful for calculations involving numbers of molecules.

A simpler way of doing the actual arithmetic is to take the concentration in ppm, multiply by the MW of water and divide by the MW of CO_2:

$$\text{Molar ratio} = 250\text{ppm} \times \frac{18}{44} = 102 \times 10^{-6} \text{ mols/mol}$$

The solubility of gases

The solubility of a gas in water depends on temperature and pressure, according to Henry's Law:

$$x = \frac{p}{H}$$

where x is the amount of gas dissolved, p is the partial pressure of the gas in contact with the water, and H the Henry's Law constant for that gas at that temperature.

This law represents an equilibrium between the number of gas molecules bombarding the liquid surface and trying to get in (measured by p) and the number of dissolved gas molecules in the water trying to get out (measured by x). The constant H shows the relationship between the rates at which these two are going in opposite directions. Rising temperature makes dissolved molecules more active but (at constant pressure) does not increase the rate at which gas bombards the water. As the temperature rises, therefore, the constant H increases and the solubility falls.

137

Reference books give tables of H for different gases at various temperatures. Table 15.1 shows H for CO_2, with values in mols/mol (as is common in text books) and in ppm CO_2 (which is handy for engineering purposes).

Table 15.1. Henry's Law constant for CO_2

$x = p/H$ where p is the partial pressure of CO_2 in bars, and x is the concentration of CO_2 dissolved in water

Temperature °C	Value of H to give x in mols/mol	Value of H to give x in ppm CO_2
0	728	0.30×10^{-3}
10	1040	0.43×10^{-3}
20	1420	0.58×10^{-3}
25	1640	0.67×10^{-3}
50	2830	1.16×10^{-3}

The composition of air

Dry air contains about 79 per cent nitrogen and 21 per cent oxygen by volume, that is by molar ratio, plus about 0.03 per cent of CO_2 and traces of other gases. The amount of water vapour is of course variable, but its partial pressure is always insignificant in degasser calculations.

The 0.03 per cent of CO_2 which is 300ppm in air at 1 bar pressure, has a partial pressure of 300×10^{-6} bar. At 10°C, according to Henry's Law, this is in equilibrium with 0.7ppm CO_2 in water. Bringing water and air into contact with one another can therefore reduce the CO_2 in solution to a very low level.

Gas-liquid mass transfer

This desirable low level will only come about if air and water are brought to near equilibrium with one another, which needs an apparatus in which the CO_2 is given the opportunity to transfer from the body of the water into the air. The urge, or driving force, which promotes this transfer is rather small. What is more, there are barriers at the surface between air and water which tend to slow down the rate of transfer. These problems are very similar to those of heat transfer. Just as in heat transfer design the degasser designer must seek to do three things: to arrange the process in such a way as to obtain the highest overall driving force; to use a contact apparatus which will allow the transfer to take place over a large area; and to reduce the barriers at the transfer interface. All these have to be achieved at the lowest cost.

Increasing the driving force

When air and water are in equilibrium there is no mass transfer because there is no driving force to promote it. The driving force only exists when the two are not in equilibrium; the greater the disequilibrium, the greater the driving force.

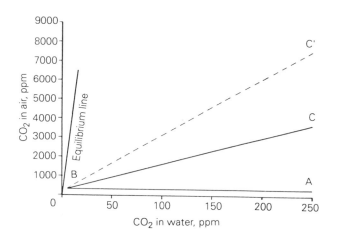

Figure 15.1. Concentrations of CO_2 in air/water

Figure 15.1 is a graph of concentrations of CO_2 in air against those of CO_2 in water. On it is drawn the equilibrium line, any point of which satisfies Henry's Law at 10°C (the normal design temperature).

Suppose we want to degas a water containing 250ppm of CO_2 to give a product of 5ppm. Point A on the graph in Figure 15.1 represents the incoming water at 250ppm, in contact with fresh air at 300ppm. This is a long way from the equilibrium line, showing that the driving force which promotes transfer of CO_2 out of the water is quite large. The CO_2 leaves the water and enriches the air, so that the CO_2 concentration in the air increases to some such point as C, which is nearer the equilibrium and therefore the driving force falls in the process. Of course, if the designer uses an infinitely large flow of air, then the actual enrichment remains negligibly small and the product water at 5ppm would still be in contact with fresh air at 300ppm, as shown on point B. During the process the water would have gone through all the points on the line AB. Such a line describing the actual course of events is called the operating line.

139

The use of an infinitely large air flow is of course uneconomical and the designer must therefore pick some limited amount of air and allow for the enrichment of CO_2 in it.

In countercurrent contact, air first contacts the water leaving the apparatus and leaves at the point at which water enters it. This means that fresh air contacts the product water with its 5ppm CO_2, as represented by point B in Figure 15.1, and then picks up CO_2 progressively as it goes through the apparatus. Suppose, for example, that we are to degas 45m³/h of water, for which we shall use 1800m³/h of air (which is a realistic sort of figure). Figure 15.2 shows the quantities involved: the air leaving the apparatus will be enriched to 3650ppm CO_2, and in Figure 15.1 this point is represented by the point C. If we ignore the small weight of CO_2 which actually passes from water to air, the operating line is a straight line connecting BC. Every point on this line corresponds to the conditions at some actual point in the degasser.

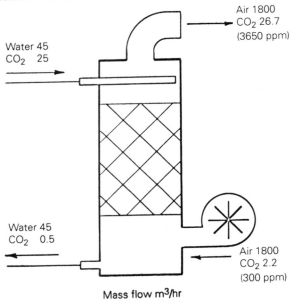

Mass flow m³/hr

Figure 15.2. Countercurrent mass balance

The operating line BC (Figure 15.1) does not provide as great a driving force as AB, but the actual difference is small. If we used less air than this, the enrichment would be greater and the operating line would rise to something like BC'. The less air used, the higher C' will end up and the nearer the operating line will get to the equilibrium line. There is a theoretical limit when so little air is used that the operating line touches or crosses the equilibrium line, because if the two lines touch then the driving force at this point is nil and this limiting condition is of academic interest.

In the example illustrated in Figure 15.1 the smallest driving force is at point B, where the water leaves the apparatus, and in degassing this is always the critical design point. If the water were wanted at, say, 2ppm CO_2 then the problem would call for a much more efficient transfer apparatus.

The transfer unit

One way of contacting a gas and a liquid is in a tower fitted with perforated plates or trays; liquid cascades down the plates while the gas flows upwards. A perfect theoretical plate would be one on which gas and liquid reach equilibrium with one another; this is a concept which is useful in measuring the difficulty of carrying out a particular transfer operation. It is possible to calculate how many theoretical plates are necessary to perform the operation under the specified circumstances. The operation is then characterized by the number of theoretical plates (NTP) which are needed. The NTP depends only on the inlet and outlet concentrations of gas and liquid, and on their physical characteristics. It is independent of the scale of operations or the type of apparatus used.

In degassing it is common to use packed towers for which the theoretical plate idea is often modified into a transfer unit (TU). In CO_2 stripping, as Figure 15.1 shows, both the equilibrium and the operating lines are straight lines, which makes it very simple to calculate the number of transfer units (NTU). Assuming a very large gas flow, which is normal,

$$NTU = \log_e \frac{x_1 - p_2/H}{x_2 - p_2/H}$$

x, p and H are the same as in Henry's Law, and the suffixes 1 and 2 refer to water inlet and outlet respectively. p_2/H actually represents the CO_2 concentration of water in equilibrium with fresh air, which we have calculated at 0.7ppm. This is usually so small when compared with x_1 that it can be ignored. Using ppm CO_2 as units, the simplified equation becomes

$$NTU = 2.3 \log_{10} \frac{x_1}{x_2 - 0.7}$$

In our example this works out to

$$2.3 \log_{10} \frac{250}{5 - 0.7} = 4.0$$

The design of packed towers needs a fair safety margin and it is usual to add at least 1TU to the calculated number.

A simpler but less reliable calculation is to assume that every halving of the CO_2 concentration in the water takes a TU. In our example the CO_2 is to be reduced by a factor of 250/5 which equals $2^{5.6}$, that is 5.6TUs, which in this case gives much the same answer as the more sophisticated formula plus safety margin.

Tower packings and the HTU

So far we have only dealt with the basic data of the process itself. The designer must also design the apparatus to make the transfer barriers as low as possible to present the greatest possible transfer area (where possible really means 'possible consistent with economy'). All kinds of tower packings are made with these objects in mind. Their manufacturers publish the properties of each type to show the mass transfer capability of the packing in different sets of circumstances. A convenient unit for this is the HTU, the height of packing in a tower which gives a performance equivalent to a theoretical plate or a transfer unit, under some given set of circumstances. Degasser towers normally use some form of dumped random packing, for which typical values of HTU lie between 150 and 600mm.

To design an actual tower it is necessary to calculate the NTU for which the problem calls, find the HTU from the packing data and multiply the two to obtain the total height of packing in the tower. The worked example above called for 5.6NTU; if we were to use a packing with an HTU of 550mm, then the total height of packing needed is

$$550 \times 5.6 = 3\text{m of packed height.}$$

Dumped packings are simple geometrical shapes; they must be corrosion resistant for CO_2 stripping. They are randomly dumped in the tower so that water dripping from one to the next is, in theory, randomly re-distributed. The classical packing is the Raschig ring, which is a section of pipe of equal length and diameter and the thinnest wall thickness consistent with mechanical strength. For corrosion resistance they are made of ceramics or plastics.

Raschig rings are cheap to make, but for their cost, weight and volume they do not present much useful surface area. They are being superseded by shapes which cost more per unit volume but have much greater efficiency. They permit greater air and/or water flow and have a lower HTU. Table 15.2 gives some typical details.

Table 15.2. Packing data

Type of packing	Pieces per litre	Weight per litre (kg)	Contact surface per litre (m^2)	Free gas space (%)
Ceramic Raschig rings 38mm × 38mm	145	6	3.6	75
Ceramic Intalox saddles, 38mm	230	7	4.6	74
Ceramic Pall rings, 38mm	150	6	4.4	76
Plastic Pall rings, 35mm	185	0.8	4.1	90

Rating of tower

The upward rush of air in the tower tends to hold up the water on the packing which is a useful effect because it makes sure the packing is thoroughly wetted. Above certain velocities, however, this hold-up becomes so great that the design flow of water cannot all get down the tower. An excess of water accumulates above the packing and the tower floods. The art of packed tower design therefore calls for the water and air flows to be matched to one another so that the tower is properly loaded without actually flooding. If too much safety margin is used, and the tower is under-loaded, then the HTU increases and the tower may actually be less efficient than it ought to be.

The whole elaborate theory of countercurrent mass transfer falls absolutely flat if the apparatus does not operate in countercurrent. This means that air must rise and water must fall with even and uniform distribution at every point in the tower. The inlet water distribution at the top of the tower is very important. Water also tends to start channelling as it goes down the packing; if the packed height is much more than five tower diameters, it may be useful to put in a re-distributor. Air can also be trapped and carried downwards by the downflow of water. This is most likely to happen when large packing size, high water velocity and low air velocity are combined; the resulting circulation of air within the tower must be avoided at all costs.

External problems

While blowing CO_2 out, a degasser can blow something else into the water. It

143

is certain, for instance, to saturate the water with oxygen in the air. This is annoying if, for example, the degasser treats a mixture of fresh makeup and condensate, in which the condensate is more or less free of oxygen, but after degassing the mixture ends up saturated with it.

Any dust or dirt in the air gets washed out by the water. For example, a degasser in a chemical factory in the English Midlands picked up so much colloidal sulphur from the works atmosphere that the anion units clogged with it. An air filter had to be installed.

Tower packings of plastic are normal today, because they are lighter and tougher. Over-enthusiastic dumping of ceramic packings used to cause trouble due to fragments of ceramic passing on with the water and causing chemical or mechanical trouble further down the line.

One source of difficulty which is usually ignored is the CO_2 concentration in the air. 300ppm is always taken as a norm for fresh air but the true value varies from day to night, summer to winter and between town and country. Readings up to 600ppm have been recorded in the country, and higher local concentrations might well occur in a large factory. 600ppm doubles the p_2/H term in the NTU calculation. With a 5ppm residual value this is still not very serious, but if, for example, we tried to design a tower to reduce 150ppm to 2ppm, the increase in NTU is noticeable:

CO_2 in fresh air (ppm)	NTU calculated (without safety margin)
300	4.7
600	5.7

Conclusions

Degassing by air blowing is a cheap way of removing alkalinity after a cation exchanger. CO_2 removal is luckily a very simple case of gas stripping, which makes the process design of towers relatively simple. The main problems lie in the correct choice of packing to make sure the tower is properly loaded, and in design of distribution system. Modern proprietary packings offer the designer a wide choice.

The process saturates water with oxygen, which may be undesirable. If the atmosphere is dirty, it may be necessary to filter the air and so avoid blowing air-borne rubbish into the water.

Values of CO_2 down to 5ppm can be obtained quite easily. To reach lower levels calls for a disproportionate amount of extra packing, and the performance of the towers becomes more critical. Amongst other factors it may be affected by the CO_2 concentration actually found in the air on the site.

16 Options in demineralization flow sheets

Demineralization plants, like any other process plant, have to meet a demand at the lowest cost. The designer's skill is only truly tested in combining these objectives, especially as the lowest cost includes both capital and running cost. Lowering the one usually raises the other, and as there is a wide choice of flow sheets selecting the best one can become quite difficult.

This section shows some of the choices in demineralization. We have worked out running costs for a few flow sheets over the normal range of water analyses, represented by three typical analyses shown in Table 16.1

Price of chemicals and water

The prices we have taken as a basis are set out in Table 16.2. We have taken 1990 prices for sulphuric acid and caustic soda bought in large bulk, so they represent the lowest prices then available.

Water used to be so cheap that the cost of waste water could be ignored. If the water comes from public supply, this is no longer true and even more so if there is a charge for disposal of effluent water.

The waste effluent water produced by a plant of a given flow sheet is roughly proportional to the TDS removed, and it is just these high-TDS waters which tend to cost the most, a clear case of Murphy's Law. In Scotland, for example, waters of about 100ppm TDS cost about 15p/m^3. In SE England you tend to pay twice as much for waters which typically contain 300ppm TDS. We have therefore taken different prices for our three waters, based on representative prices for 1990.

We have not costed the extra cost of effluent disposal. As stated, it tends to be higher in areas with high-TDS water, and can raise the effective price of water by as much as 100 per cent.

Table 16.1. Typical water analyses (ppm CaCO₃)

	A	B	C
TDS	80	250	400
Hardness	60	200	320
EMA	55	50	200
Alkalinity	25	200	200

Table 16.2. Assumed chemical and water costs

H_2SO_4 £100/tonne 100%

NaOH £250/tonne 100%

Waste water:
Analysis A $15p/m^3$
Analysis B $25p/m^3$
Analysis C $35p/m^3$

The flow sheets

Table 16.3 which shows five flow sheets numbered FS1–5 is the real meat of this chapter, and all the rest which follows is a commentary on it. Room has been made in the table for each reader to calculate his costs using his own prices of chemicals and water with our consumptions.

For each flow sheet we have worked out running costs on our assumed data. For simplicity we have only worked on the basis of co-flow regeneration with H_2SO_4 without effluent neutralization. This leaves a lot of alternatives which are discussed in the text.

Raising the regeneration level of a resin increases its capacity at the cost of reduced chemical efficiency, that is it cuts capital cost and increases running cost. We have taken middling regeneration levels as a basis for our costs.

Capital costs

Capital costs vary so much that there is no way of showing them numerically. Generally speaking, flow sheets FS1–5 are arranged in increasing order of capital cost, and the range between them might be as much as 1:3; the order can be upset if a flow sheet is applied to an unsuitable water. For example on Analysis B, FS2 will cost more than FS3 (see p.148).

Table 16.3 shows that, while capital costs rise from FS1–5, running costs fall. At the same time the flow sheets are also arranged in steps of rising water quality (see below). FS1 which leaves all the SiO_2 and some CO_2 in the water is in a special category in this respect. The other four flow sheets are sufficiently similar that if they are followed by a polishing mixed bed, the final treated water quality will be the same. Such a mixed bed costs very little to run but quite a lot in capital cost, and it needs more careful operation than simple cation or anion units. (See Chapter 17).

Treated water quality

As we have already said, FS1 leaves silica and CO_2 in the water. These can be taken out on a mixed bed, but then the whole train would depend on one silica-removal stage and could not be guaranteed to give very high quality at all times.

All the other flow sheets remove silica and CO_2 but FS4 and 5 will give slightly lower silica than 2 and 3 because they use the stronger type 1 anion resin. Most of the leakage from these trains is sodium. The leakage level depends theoretically on the raw water composition and the regeneration level of the cation resin, and is usually below 5ppm. The silica leakage from FS2 and 3 may be up to 0.5ppm, and from FS4 and 5 perhaps half that. In practice these leakages depend as much on the care with which the plant is designed and operated as on theoretical considerations. A mixed bed polisher can take out all these leakages to produce very high quality water.

Counterflow regeneration (CFR)

As FS1 is only used to produce moderate quality water it does not consider this alternative, which can be applied to the cation resin and possibly also the anion resin in all five flow sheets.

On the other four flow sheets, CFR is used primarily to reduce leakages, sodium to less than 0.5ppm, and silica less than 0.1ppm. With only a single silica-removing stage, there is always a chance of high leakages for short periods especially at the start and end of a cycle, but for many purposes the water quality from a CFR plant without a polishing mixed bed is good enough.

With HCl regeneration, CFR also improves the chemical efficiency, but with H_2SO_4 the efficiency is much the same as for co-flow at the same regeneration level. On the other hand, the low leakage from CFR plant makes it possible to raise the efficiency by designing at lower regeneration levels. CFR also wastes less water in rinsing the resins. However, CFR plant is more expensive and more complex than co-flow plant of the same size.

The flow sheets and their properties

In practice, FS3 is the commonest and therefore serves as a starting point in

Table 16.3. Running costs resulting from selected process flow sheets

Flow sheet	ANALYSIS	H_2SO_4 (100%) g/m³ water	NaOH (100%) g/m³ water	Waste water, %	Cost per m³ treated water		
					Chemicals p/m³	Waste water p/m³	Total p/m³
FS1							
Cation — Weak base anion — Degasser	A	196	77	3.5	3.8	0.5	4.3
	B	612	70	8.4	7.9	2.1	10.0
	C	980	280	15.7	16.8	5.5	22.3
FS2							
Cation — Type II Strong base anion	A	196	173	4.6	6.3	0.7	7.0
	B	612	540	14.2	19.6	3.6	23.2
	C	980	864	22.8	31.4	8.0	39.9
FS3							
Cation — Type II Strong base anion — Degasser	A	196	119	4.0	4.9	0.6	5.5
	B	612	108	8.8	8.8	2.2	11.0
	C	980	432	17.5	20.6	6.1	26.7
FS4							
Cation — Weak base anion — Type I Strong base anion — Degasser	A	196	149	4.4	5.7	0.7	6.4
	B	612	142	9.4	9.7	2.4	12.1
	C	980	352	16.5	18.6	5.8	24.4
FS5							
Weakly acidic cation — Cation — Weak base anion — Type I Strong base anion — Degasser	A	188	149	4.0	5.6	0.6	6.2
	B	362	142	7.7	7.2	1.9	9.1
	C	803	352	13.2	16.8	4.6	21.2

design thinking. In some but not all circumstances it provides the right answer for all our analyses.

Where silica and CO_2 may remain in the treated water, then the high capacity and efficiency of the weak base resin in FS1 gives the lowest capital and running costs; the advantage is much reduced if the effluent has to be neutralized (see p.150). FS2 stands alone in removing all the anions by ion exchange, bicarbonates and all. This gets expensive if the bicarbonate load is significant, not only from the extra running cost of regenerant but because the high anion load may mean that a larger volume of anion resin has to be installed which cancels the saving from omitting the degasser.

Anion resins need a minimum contact time to work properly, and a minimum resin volume must be used for a given flow, which may lead to long cycle times. On a very thin water we may need so much resin that it might as well cope with the bicarbonate, and if the effluent has to be neutralized the running cost penalty can be quite small, too. If these circumstances come together, this very simple flow sheet can be used economically.

FS4 and 5 represent attempts to get lower running costs than FS3, though the extra complication usually means higher capital costs. In FS4, weak base resin takes out the EMA using less caustic than the strong base resin of FS3; if the EMA is high (as in Analysis C) there can be an overall advantage. In FS5, the weakly acidic resin uses only a little excess acid in taking out the temporary hardness cations. We would therefore expect FS5 to pay off on high-alkalinity waters such as Analysis B but, as acid is relatively cheap, the potential running cost saving is small. Taking out the temporary hardness cations yields a water with an increased percentage of Na^+ which then has to be treated by the strong cation resin. To keep the Na^+ leakage down then requires either a high (and inefficient) regeneration level, or CFR (see Chapter 14). (For moderate quality water and smaller outputs there is a flow sheet, not shown in Table 16.3 which overcomes this Na^+ leakage problem: weakly acidic/degas/mixed bed.)

The influence of raw water analysis

Analysis A is a low-TDS, low alkalinity water, typical of hilly places like Scotland. Its low TDS means that chemical and water consumptions are low for all flow sheets with little scope for cutting running costs. For the silica-removing flow sheets FS2–5 the highest running cost is only $1.5p/m^3$ more than the lowest.

The alkalinity is so low that FS2 might prove right, rather than FS3, but FS4 and 5 are quite unsuitable. The kinetic limitations on anion resins, mentioned above, force us to put in relatively large resin volumes with long cycle times, and if we split the anion load between a strong and a weak base resin, as in

FS4 and 5, then we find ourselves using large resin volumes for a very small cut in running cost.

Analysis B has a high alkalinity and is probably a deep-well water from chalk or limestone country. Here a degasser is essential for economy, but after a degasser the anion load is actually smaller than in Analysis A, so a weak base resin as in FS4 and 5 is even more unsuitable. On the other hand, comparing the running costs of FS4 and 5 shows that there is a small economy to be had from the use of weakly acidic resin.

Analysis C might well be a surface water whose TDS is partly due to previous use and discharge, such as the Thames at the London end. It has the highest TDS and EMA, and therefore the highest running cost and the most scope for economy, shown by the progressive fall in costs from FS2 to FS5; not that anyone would dream of using FS2 on the large industrial scale.

Series regeneration

Our sums assume that in FS4 and 5 each resin is regenerated independently. Alternatively we can pass the regenerant first through the strong resin, and then use the excess in the part-used solution to regenerate the weak resin. This leads to a significant cut in chemical cost, but it brings its problems: it calls for a very careful design to balance the performance of the two resins, and reduces flexibility if the raw water quality varies; there is the difficulty that the part-used NaOH from the strong base is loaded with silica which may precipitate when it becomes neutralized by the weak base resin; and in the same way, sulphuric acid regeneration of cation resins leads to very great danger of $CaSO_4$ precipitating in the weakly acidic resin. There is no such problem with HCl series regeneration, and the practice is therefore more familiar in the rest of Europe, where HCl is cheaper and would normally be used anyway. With series regeneration it is also possible to install each pair of resins in a single shell, which cuts capital cost.

Neutral effluent

This frequent requirement completely distorts the economics suggested by Table 16.3.

Cation resins normally use their regenerant more efficiently than anion resins. In FS2 therefore, where cation and anion loads are roughly equal, there will be more waste caustic than waste acid. But FS2 is not in common use. Most large plants have a degasser which reduces the anion load; for example in Analysis B it becomes one-fifth of the cation load. In most of those flow sheets which have degassers the bulked effluent would contain an excess of acid and need caustic for neutralization. Improving the efficiency of anion regeneration, as in FS4, then becomes quite pointless; in order to neutralize the effluent we need more waste caustic, not less. In the same way, FS1 loses much of its attraction, as we said above.

Table 16.3 shows that the economy of FS5 is only a little better than FS4, because acid is relatively cheap. But where we have insufficient waste caustic to neutralize the cation effluent, then the acid economy becomes vital not for the money saved in acid but for the saving in neutralizing with caustic, and FS5 offers a potential solution.

Future trends

The chemical consumption of strong cation and anion resins in co-flow, which is the basis of Table 16.3, lies between 2.5 and 3.5 times the minimum theoretically required. (Weakly acidic and basic resins work at much higher efficiencies.) If we work out the theoretical minimum chemical consumption together with our assumed cost data, we get the following results:

Raw water analysis	A	B	C
Theoretical cost of chemicals	2.2	7.5	9.5 p/m^3

There is therefore quite a lot of scope for reducing running cost.

If we consider FS5, complete with series CFR regeneration, we have reached a sort of ultimate in fixed-bed demineralization design. Plants are in fact being built on these lines and are claimed to work at chemical consumptions as low as 1.15 times theoretical. Another possibility, which is gaining ground for high-TDS waters, is to treat the water first with RO or ED to remove most of the dissolved solids. This yields an intermediate product of quality A or better, and has the advantage of putting a smaller chemical load on the environment. The main problem is the pretreatment often required to avoid membrane fouling, so capital costs often come out very high. Whether this high sophistication is worthwhile depends on a great many factors, for example: the extra capital cost and the availability of capital; the probable utilization of the plant; the operational flexibility needed. The designer's job is to balance these mutually exclusive requirements and find an overall optimum.

17 Mixed beds

We have described how, by successive cation and anion exchange, a water can be completely demineralized. We showed that these two steps have to be repeated at least once to obtain a product of high purity. However, if the two resins are mixed, so that cation and anion exchange are repeated many times, then a product of the very highest purity results. The mixed-bed process is therefore the best suited for the production of high-grade water, but the design and operation of mixed beds present chemical as well as mechanical problems.

The mixed-bed process

The mixed-bed cycle depends on separating the two resins so that they can be regenerated individually and then remixing them for the treatment of water. Anion resin has lower density than cation resin and a mixture of the two resins can therefore be separated by upflow classification, that is by backwashing. After regeneration in two distinct layers one on top of the other, the two resins are re-mixed mechanically and the mixed bed can be put back into service.

A typical mixed bed is shown in Figure 17.1. It is basically similar to a conventional ion exchange unit: a cylindrical vessel contains a resin bed, above which there is the usual free space to allow the bed to expand when it is being backwashed. In addition to the usual distributors, a mixed bed is fitted with a centre distributor.

When the resin is to be regenerated, the bed is first backwashed. This causes it to expand and allows the heavier cation resin to sink to the bottom while the lighter anion rises to the top (Figure 17.1a). After a short time on backwash this separation is complete and the upflow is stopped. The resins then settle without upsetting the separation, and there is a well-defined interface between the bed of cation and the bed of anion resin.

The unit is so designed and the resin quantities so calculated that this interface is just at the level at which the centre distributor is fixed. This distributor is fitted with mesh-covered holes or with nozzle strainers so that water can pass but resin is retained. It makes it possible to regenerate the anion resin with caustic soda and to rinse it, withdrawing the spent solutions through the centre distributor and without involving the cation resin beneath (Figure 17.1b). After this operation is complete, the cation resin is similarly regenerated and rinsed without involving the anion (Figure 17.1c).

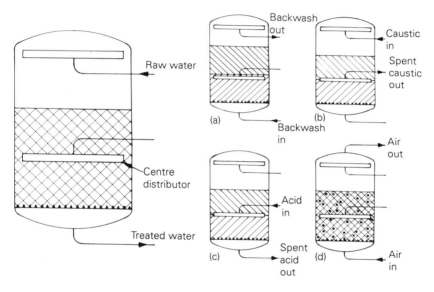

Figure 17.1. Typical mixed-bed unit operation

When both resins have been regenerated and rinsed, the excess water above the bed is drained down to the surface of the bed, after which air is blown in through the bottom distributor and out of the air release at the top of the unit (Figure 17.1d). The mechanical agitation which this causes will re-mix the resins. When the resins are properly mixed, the space above the bed is re-filled from above in order to avoid disturbing the mixed resin, after which the unit is ready to be given a final rinse down before being returned to service.

During air mixing, the small residues of regenerant come into intimate contact with both resins. In theory, the cation resin should remove any NaOH left on the anion resin and the anion resin any HCl or H_2SO_4 left on the cation resin, so that the water in which the mixed resins lie ought to be completely demineralized during air mix. In theory, therefore, a well-regenerated and well-mixed bed needs no final rinse at all. In practice, however, none of this is quite so easy.

Mechanical problems

One immediate problem in design is the physical strength of the centre distributor. A packed resin bed on downflow has very high internal friction, and a distributor buried in it is more or less locked into the resin mass. If the bed moves, it will transmit very great strain to the distributor. Resins inevitably swell and shrink as they change ionic form, and during service there is also a tendency for the bed to compact. This means that the distributor has to withstand a distributed load equivalent to the hydraulic pressure exerted by the incoming water. A 2m diameter bed operating at 0.5 bar pressure loss represents a very moderate condition, but even this is equivalent to a load of 15 tonnes on the distributor. During air mix, especially when the air flow first goes up through the settled bed, there may be a tendency for large bubbles of air to come up one side of the vessel, pushing large masses of agglomerated resin with them. This exerts heavy torsional forces in an upward direction.

Hydraulically the distributor has to be designed to spread the regenerants evenly across the cross-section of the bed. This means that the liquid outlets must be distributed across the bed by some configuration like a header-and-lateral system, which is going to obscure a high portion of the cross-section. It is possible to take this out of the centre of the bed, where it is exposed to such large forces, by using a design which has a header-and-lateral system in the water space above the bed, with vertical downcomers to each liquid distribution point. The downcomers have to be much longer than they do for the CFR design illustrated in Chapter 14. The device is used to some extent in Europe, but is rather an expensive expedient.

The distributor has to resist attack by both acid and alkali so that expensive or weak corrosion-resisting materials have to be used; this aggravates the problems of design.

In some applications, the mixed bed treats water which is already very pure. In condensate polishing, for example, the chemical load put on the mixed bed is so small that the exhaustion run can go on for weeks. The chemical kinetics of mixed beds are so good that very high liquid velocities can be used, reducing the size of the mixed bed and therefore its capital cost. Actual flow velocities go as high as 150m/h ($50g/ft^2/min$). Such a flow rate raises large pressure losses and any centre distributor which was strong enough to withstand the resulting force would cause unacceptable flow disturbance within the bed. In these cases it is now normal practice to build a separate regeneration plant and to move the resins out of the mixed bed so that they can be separated and regenerated in equipment which is intended specifically to carry out these duties.

Problems of separation

The basic requirement for good mixed-bed operation is that the resins must remain intimately mixed during exhaustion and totally separated during regeneration. The more we fail to achieve this ideal state of affairs, the less good will be the product water and the chemical efficiency of the unit.

In fact, separation by backwash is never perfect, and some traces of each resin remain in the wrong bed.

There is a very interesting problem with brand-new resins in mixed beds, which occasionally refuse to separate at all. An ion exchange resin is described as a lattice work with charged groups fixed on it, and each of these charged groups attracts ions of the opposite charge. There is no reason why one ion exchanger should not attract another ion exchanger of the opposite sign; in fact the attraction between them is extremely strong. Mixed beds of new resins, especially in the regenerated form, will tend to cluster together unless the electrostatic forces which hold them together can be broken. This is quite easily done by dousing them in strong solutions. Within a very few cycles, the resins have coated themselves with a surface film of colloidal particles or macro-ions and this de-activates the surface of the resin to such an extent that the problem disappears.

Even when the electrostatic clustering problem is out of the way, separation by classification is never perfect and some traces of each resin remain, as stated, in the wrong bed. This inevitable imperfection is aggravated by air bubbles which may adhere to cation beads and carry them up into the anion resin. Organic contaminants in water tend to build up on anion exchangers, not only on the resin surface, as mentioned above; they also migrate slowly into the body of the bead and accumulate there. This accumulation interferes with ion exchange in many ways, one of which is that the accumulated material in the resin actually increases its density. An old and badly fouled anion resin will separate badly from the cation resin because of the inadequate density difference between the two. Classification depends on differences in the terminal velocity with which single resin beads fall in water; this in turn depends on both the size and the density of the resin. Special grades of resin should be used for mixed beds in order to ensure that the particle sizes are, as near as possible, uniform and that the particles will therefore separate by virtue of their density difference alone. With use, however, the resin tends to break up and yield a proportion of fines. Some of the cation fines will then classify with the anion resin which is lighter but bigger. There is no way of avoiding this except by renewing the resin charge.

Good separation is only meaningful if the centre distributor is exactly at the interface between the two resins. This cannot be guaranteed. The interface level varies slightly between cycles because the bed may compact to a greater or lesser extent. Loss of resin by attrition reduces the resin volume and in time causes the interface to fall below the distributor level, unless of course the designer has allowed for some attrition by setting the interface above the distributor level when the charge is first put in.

To sum up, therefore, there are three ways in which resin ends up in the wrong part of the bed: by imperfect classification; by cation fines classifying with the anion; and by failure of the centre distributor to match the interface level. Good design and operation of the plant can minimize the amounts involved but it cannot totally eliminate any of the three.

Problems of mixing

The main variables in mixing are the air flow used and the amount of water remaining behind after the rising space has been drained down. The residual water is critical. If it is drained to leave a few inches of water above the top of the bed, then air agitation produces a sloppy fluid of low viscosity in which a good mix can quickly be obtained. As soon as air mixing is stopped, however, the bed is left suspended in an excess of water and has to settle to the bottom of the unit. As it does this, the heavier cation particles slip to the bottom, leaving the lighter anion behind. The falling resin displaces water from the bottom of the unit, so that there is an actual upward flow of water within the bed, and although the bed may only have to fall a few inches, there is a surprisingly rapid partial re-separation of the resins. Thus the good work of mixing is undone. Some operators open the bottom drain as soon as the air flow is stopped, causing a downflow of water which helps to overcome this effect.

If, on the other hand, too much water is drained off before mixing, then there will not be enough left behind to provide lubrication between the particles; the mix is then stiff and the resins will not mix properly. With a stiff mix the air flow is much more critical. Too much air tends to blow all the interstitial water out of the bottom of the bed, so that an immovable, semi-dry section forms, through which the air passes quite easily without causing the resins in it to mix. The water thus blown out of the bottom of the bed then helps to make the top of the bed a little sloppier. Mixed beds are often fitted with sight glasses at the interface level, but an appearance of good mixing at that level is no guarantee that the bottom of the bed is moving freely.

Every operator who regenerates mixed beds manually develops a pet routine on the best way of doing it. Some like to mix in two stages: sloppy first and finally stiff; some like to bleed water from the bed during mixing so that the mixture stiffens gradually (and in so doing run the danger of over-draining and air-locking the bed); the real experts mix by ear, and can distinguish between the 'slurp-slurp' of a sloppy mixture and the 'glup-glup' of a stiff one. These games are relatively easily played on small plants with manual control and good windows; they become much more difficult on large or automatic plants. There are, however, useful devices which can be built into the automation control to bring about a consistently good air mix. Even so, most mixed beds probably end up with a small layer of cation resin at the bottom of the unit.

Chemical effects: bad mixing and separation

Inevitably, then, some wrong resin remains in both resin beds during regeneration, which is then exposed to the wrong regenerant. As a result, these resin fractions are not merely exhausted but return to the next service stage totally saturated.

In normal cation exchange, for example, the cation resin starts out regenerated partially to the H^+ form and ends up in equilibrium with the raw water at about pH7. As the raw water contains a concentration of 10^{-7} of H^+ ions (see Chapters 9–13) this equilibrium means that a few H^+ ions must also remain on the resin to represent the minute number of H^+ ions which actually manage to fight their way back from the water onto the resin. But the fraction of cation resin which is caught up in the anion bed is treated with a caustic soda solution at pH above 12, with an H^+ concentration of less than 10^{-12}, (i.e. practically nil). The resin accordingly picks up more Na^+ ions and ends up with no H^+ ions on it at all. When this saturated fraction of cation resin is once more brought into contact with water at pH7 or thereabouts, the few H^+ ions which now get back on to the resin from the water will make the resin bleed an equivalent number of Na^+ ions into the water.

Strongly acidic resins have such a high affinity for Na^+ compared with H^+ that the number of H^+ ions which water at pH7 puts back on them is minute, and therefore the effect of the saturated cation resin fraction is unimportant. The anion resin which has ended up in the wrong bed is exposed to acid, causing the equivalent but opposite effect. The acid saturates the resin with Cl^- (if HCl is used) and when brought into contact with water at pH7 the resin then bleeds Cl^- because a few OH^- ions are taken up from the water. Unfortunately, anion resins are less ideal than cation resins. The strong base resin's active group (a quaternary ammonium group) is prone to chemical degradation to a form which is still an anion exchanger but much less strong than the quaternary ammonium. On these degraded weak base groups the affinity

157

difference between Cl^- and OH^- is much less great and so the bleed which follows saturation with Cl^- becomes much more serious.

Degradation of the quaternary ammonium group is the result of fouling, oxidation and just old age. To some extent degradation actually takes place during manufacture of the resin, so that even resins which are made under conditions of strict quality control already contain significant proportions of weak base groups when new. The Cl^- bleed from the anion resin in a mixed bed is therefore always greater than that of Na^+ from the cation resin. The exchange of Cl^- ions for OH^- lowers the OH^- concentration and therefore lowers the pH. Waters coming from mixed beds therefore tend to be acid rather than alkaline. The increased H^+ concentration at the lower pH does however also release a very small extra quantity of Na^+ into the product.

The quality of a water from an ion exchange column is always dictated by the condition of the last section of the bed, with which the water will very nearly be in equilibrium. A mixed bed is homogeneous after air mixing so that some saturated anion resin must be at the bottom of the bed. The actual mass of Cl^- which it bleeds into the product water is very small, but as mixed beds are used to obtain very high purities, the bleed inevitably shows up in the product. Moreover, mixed beds are usually monitored by conductivity and an acid leakage, such as we expect from this type of bleed, raises four times the conductivity which would result from an equivalent amount of neutral salt (see Figure 17.2). As little as 0.05ppm as $CaCO_3$ of HCl will give a conductivity of $0.4\ \mu S$.

A good mixed bed needs practically no final rinse. Depending on the efficiency of the operation and the condition of the anion resin, mixed beds in practice may require up to half an hour of rinse in order to remove the bleed of anions, and in really bad conditions it may be impossible to get the conductivity down at all.

This bleed, which is a chemical phenomenon, must not be confused with licking, which is the result of stagnant pools of regenerant left behind at the bottom of the unit after regeneration. Licking is a common problem in badly designed single resin units but in mixed beds the air mix makes its occurrence unlikely.

Chemical effects: matching the capacities

The chemical process by which ions are removed in a mixed bed looks very simple, but only at first sight. In actual operation it is quite complicated.

Consider, for example, a mixed bed which treats a solution of 20ppm as $CaCO_3$ of NaCl, in which the resins are mixed in equal volumes. Suppose the cation and anion resins have been regenerated to give them a working capacity of 640 and 460meq/l respectively. 20ppm NaCl is equivalent to 0.4 meq/l.

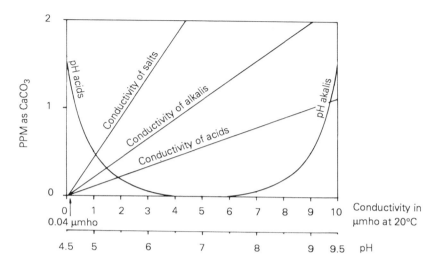

Note: It is important to remember that pH measurement in highly demineralized water is difficult and unreliable because of the low conductivity of the water itself, and also because minute pick-up of impurities will cause large errors.

Figure 17.2. Effect of leakage on pH and conductivity

Table 17.1. Units of measurement, conversion table

ppm* CaCO$_3$	meq/l†	kgrn/ft^3 CaCO$_3$	lb CaCO$_3$/ft^3
1	0.02	0.436×10^{-3}	62.3×10^{-6}
50.0	1	21.8×10^{-3}	3.11×10^{-3}
2290	45.8	1	0.143
16 100	321	7.0	1

* Or mg/l CaCO$_3$ † or epm or mval/l NB 1 meq/l $\equiv$ N/1000

Note: The column showing lb CaCO$_3$/ft^3 is added to assist in calculation of regenerant doses.

If the two resins were separate, then it is clear that a litre of cation resin could treat 1600 litres of water before becoming exhausted, and a litre of anion resin would treat 1150 litres. Mixing the resins does not in itself change their capacities. If we mix a litre of each to give two litres of mixed resin, then after 1150 litres of water have passed through the mixture the anion resin in it will be exhausted, while the cation still has capacity left until 1600 litres have passed through the bed. What happens is that the first drops of water react with both cation and anion resin at the top of the regenerated bed, and then an anion exhaustion zone and a cation exhaustion zone form. These two zones are independent of one another, and as the anion resin is more quickly exhausted, the anion exhaustion zone moves down the column more quickly and leaves the cation zone to follow behind.

Figure 17.3 shows this effect. It represents the mixed bed described above when it is about half-way through its service run. At the top of the bed zone A contains exhausted resin of both kinds through which the raw water passes unchanged. At the top of zone B is the cation exchange zone, but in zone B the anion resin is already exhausted so that this zone behaves like a cation resin column. At the top of zone C, in which both resins still have capacity, is the anion exchange zone. It is only in zone C that true mixed-bed conditions exist; it is only here that the water is subjected to repeated cation and anion exchange.

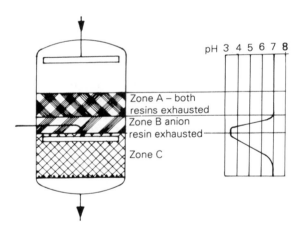

Figure 17.3. Exhaustion of mixed-bed unit with low anion resin component

The curve at the side of Figure 17.3 shows the pH value in these zones. In this example the pH in zone B falls to about 3.4, and then rises (theoretically) to 7.0 in zone C.

The above example took a resin mix in which the anion resin's capacity limited the capacity of the bed. By increasing the proportion of the anion resin and by regenerating it to give a higher capacity it is possible to change this. Suppose, for example, the resins were regenerated to the same capacities but were in the proportion 1:2 cation to anion. A two-litre bed will now have the capacities due to ⅔ litre of cation and a 1⅓ litre anion, which are 1067 litres and 1533 litres of water respectively. In this mixed bed, the cation exhaustion zone will run ahead of the anion exhaustion zone, and in the intermediate zone B under these conditions the pH will run up to 10.6. Figure 17.4 shows this kind of operation diagrammatically.

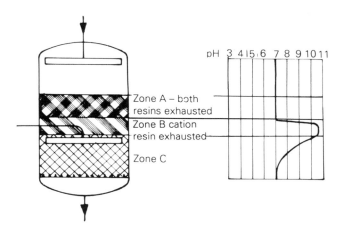

Figure 17.4. Exhaustion of mixed-bed unit with low cation resin component

As soon as one or other of the exchange zones reaches the bottom of the bed, the unit must be regenerated, and any remaining capacity on the other resin is then wasted. From that point of view, the ideal design would match the resin mix and capacities exactly to the raw water analysis so that the two exhaustion zones run down the column together. In practice this is impossible because of slight variations in the capacities from cycle to cycle, and also because changes in the raw water composition must be expected. Moreover, the designer may particularly want to avoid an acid condition in zone B. Silica removal is the main duty of many mixed beds and silica is so weakly dissociated that in acid solution it is completely suppressed. Worse, the silica may actually go into a polymeric form which is not properly removed when the water gets into zone C, and which may build up on the resin as a foulant. For this type of duty it is important to design with cation capacity limiting the mixed-bed capacity, and therefore an alkaline zone B.

161

Design of mixed beds: capacity

Compared with units containing only one resin, the capacity of a mixed bed is always low. Imperfect separation, discussed above, leads to waste as a result of chemicals contacting the wrong resin. The need to regenerate when only one of the resins is exhausted wastes available capacity on the other. As a result, the effective capacity of the resins must be down-rated by a factor of the order of 10–40 per cent below the normally expected value.

For a given capacity, therefore, mixed beds actually need more resin than would be needed in separate cation and anion units. From a capacity point of view, a 1m diameter mixed bed has less than half the capacity of a pair of 1m diameter cation/anion units. On the other hand the kinetics of mixed-bed exchange are good, so that a greater flow of water per volume of resin can be treated, which leads to shorter cycle times and a more intensive use of the resin.

Raw water analysis

Hard water cannot be treated satisfactorily in mixed beds. When, inevitably, some cation resin contacts the caustic soda regenerant, the strong NaOH solution regenerates the hardness off the cation resin and displaces it with Na^+. As soon as the Ca^{++} and Mg^{++} get into solution the high pH there causes them to precipitate and clog the bed.

In a train of cation and anion units, it is possible to place a degasser tower between the units to blow out the CO_2 arising from the bicarbonate initially in the water. With waters of high alkalinity this leads to a considerable cut in the load on the anion unit, resulting in both capital and running cost reduction. A mixed bed removes CO_2 and HCO_3 by ion exchange on the anion resin; it therefore needs extra resin to provide the necessary capacity and more caustic for regeneration.

Organic macro-molecules, such as the commoner detergents or humic and fulvic acids, are always troublesome to ion exchangers. In the mixed bed there is the additional problem that fulvic acids precipitate in an acid environment; this aggravates fouling. Some of these large molecules are slow to move both in water and resin, and their removal therefore takes time. With high flow rates the necessary time is not available and the organics leak through, causing a high conductivity in the product water. Indeed, there are some organic anions too big to be absorbed at all by the usual ion exchange resins, and from waters containing these it is impossible to get very low conductivity water by ion exchange. Mixed and single beds are at a similar disadvantage in these cases but, as it is normal to use mixed beds to obtain very low conductivity, this problem is usually associated with mixed beds.

Summary: when to use mixed beds

The mixed-bed process is capable of giving very high water purity. It will treat hard water only with difficulty. Its chemical efficiency is lower than separate resin beds, especially for treating water of high alkalinity. Its capital cost is high unless it can be used at high flow rate which permits the resins to be used more intensively, and this is the case only if the incoming water is already fairly pure.

The ideal duty for a mixed bed is therefore as a polisher, either after an ion exchange train of single resin beds, or for distillate or condensate polishing. Where a very small flow of high-purity water is wanted, mixed beds are sometimes used to treat a normal potable water direct. Sometimes this is done for convenience rather than for paper economy but as the operation of mixed beds is trickier than that of single beds there are occasions when even this convenience turns out to be illusory.

The theory of backwashing

The flow of water exerts a force on every particle in a bed. In downflow the particles are packed against one another and cannot move, but in upflow the bed expands, if space is allowed for it to do so, because the force tends to lift the particles. The force exerted on each particle by the flow of water is a function of the particle cross-sectional area. The force of gravity, which is a function of the particle's weight, opposes the lift. A particle in free fall through water will accelerate until these two forces are equal, and it then continues to fall at its terminal velocity.

A bed with backwash passing upwards through it will also arrive at an equilibrium when these two forces are equal and opposite. Provided the upflow is fast enough to lift the particles, the bed expands and in so doing it increases the cross-sectional area through which water is flowing so that the actual flow velocity is reduced. The bed therefore expands until its particles are suspended at their terminal velocities.

At every cross-section of the bed there is one water velocity, but there may be particles of different sizes and therefore different terminal velocities. Large particles of high terminal velocity will sink because the water does not exert enough force to keep them up, while small particles will be washed upwards. Theoretically the bed ends up perfectly classified, with the smallest particles at the top and the biggest at the bottom.

If all the particles in the bed are of the same material, the force exerted by gravity is a function only of their volume, which is a cube of the diameter, as compared with the force exerted by the water, which is a function of the square of the diameter. Big particles therefore have a higher terminal velocity.

With two different materials, the force of gravity depends on the volume times the effective density. Although cation and anion resins have similar densities (1.27 and 1.07 respectively), their effective densities in water, on Archimedes' Principle, are 0.27 and 0.07 respectively, a substantial difference and quite sufficient to get a clean separation between the two resins even if both contain particles of different sizes. Obviously there is a limit to the biggest anion beads and smallest cation beads which can be tolerated, so some care has always to be taken in selecting resins. The effective density principle leads to the use of brine for carrying out very difficult separations. With a density greater than 1.00, the difference in effective density is enhanced.

If spheres of uniform diameter are shaken down and allowed to pack as closely as they can, they end up in a regular geometrical pattern called close hexagonal packing which has a void volume of 26 per cent. This percentage void volume is the same regardless of the diameter of the spheres, provided they are all the same diameter. However, if the spheres are of mixed sizes, then the small ones fill the spaces between the big ones and a smaller void volume can be obtained.

Technology advance

In recent years there has been an interesting addition to mixed-bed technology. One problem mentioned above is that it is in practice impossible to make certain that the centre distributor is exactly at the interface between cation and anion resins. If the interface lies higher than the distributor, the caustic soda regenerating the anion resin layer will finally pass through some of the cation resin and totally exhaust it to the sodium form. If the interface is below the distributor, the regeneration acid will pass through some of the anion resin and exhaust it to the chloride or sulphate form, depending which acid is used.

Either way, a fraction of one resin or the other will end up completely exhausted immediately after regeneration. This can affect the treated water quality — though in normal demineralization the effect is only noticeable if the very highest treated water quality is required. In condensate polishing applications it can lead to additional problems. (Chapter 23 describes the problems peculiar to condensate polishing.)

Some ion exchange resin manufacturers now make an inert resin with a density just between that of special matching cation and anion resins. When a mixed bed containing this inert resin is backwashed to separate the resins, the inert material will classify in a band between cation and anion resins. Normal practice is to add enough inert material to form a band about 150mm deep, with the levels so arranged that the centre distributor lies within the inert resin band. Figure 17.5 shows such a mixed bed during regeneration.

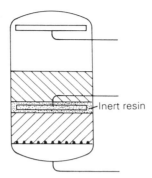

Inert resin

Figure 17.5.   Use of inert resin in mixed-bed unit

The use of inert resin thus avoids the problem of the centre distributor at the wrong level and of regenerant cross-contaminating one or other resin. There will always be some cross-contamination due to fine particles of cation resin classifying with the lighter anion resin and becoming exhausted with sodium, but with care the actual amount of cation fines can be made very small.

The inert resins are not cheap, and their use will therefore be limited to those applications where the treated water quality is critical. In practice this means that most applications of the inert resin in mixed beds will be found in condensate polishing.

18 A special use for ion exchange resin: nitrate removal

The treatment of water for potable purposes has not in the past included ion exchange as a major process. This is because drinking water does not need to be free of mineral content; indeed, minerals are necessary to make water palatable and supply part of our nutritional intake. Potable water does, of course, have to be free of disease-carrying organisms, and it should look palatable. Treatments to achieve these requirements include filtration, pH correction and a degree of disinfection. These treatments are far simpler than those used in industrial water treatment but are used on a massive scale. In some hard-water areas, base exchange softening has in earlier years been employed prior to blending to reduce the scaling characteristics of a mains supply water where this is hard, but there is some disputed evidence showing a correlation between softened water and heart disease. In general, then, the use of ion exchange resins in this market has been a minor activity. Times are changing.

Over the past few decades, medical evidence has been gathering which suggests that limits should be placed upon many of the constituents found in raw water for potable purposes. Many of the limits can be met without overmuch difficulty but in some cases we have, by environmental indifference in the past, created problems which will be with us for many decades to come. Nitrate levels are such a problem, and this chapter demonstrates ways in which the problem is being tackled, with particular reference to ion exchange.

We face ever-increasing levels of nitrate in some of our potable supplies. These are due to agricultural practice, and, since both water and nitrate can spend decades underground, cannot be reversed in the foreseeable future.

Nitrate is a potential health hazard, though this is still quantitatively debatable. Consequently, the World Health Organisation, the US government and the EC all limit the permissible nitrate level in potable water to 50mg/l as NO_3, or thereabouts.

In many affected areas the possibility of maintaining the nitrate contamination below the specified level by mixing high-nitrate supplies with those of better quality has now been utilized to the full. Many suppliers are obliged, therefore, to resort to denitrification; this involves installing processes which are very different from those historically used in the treatment of potable water.

Of many routes which have been examined, only two processes have so far emerged as suitable for the large scale: biological denitrification and ion exchange. In the biological process the water is passed through a sophisticated version of the septic tank, in which bacteria are deprived of oxygen with which to support life. They resort instead to breaking down the nitrate NO_3 to use its oxygen and liberate nitrogen gas.

In these circumstances the bacterial culture is easily upset. The process is difficult to control and, if it goes wrong, can create a greater threat to health than it prevents, by the generation of harmful nitrites or by generating undesirable strains of bacteria. This difficulty means that it needs very high-class supervision and is unlikely to come into early use except perhaps on very large waterworks where this supervision can be guaranteed. On the other hand, unlike ion exchange, it destroys the nitrate and therefore does not create a serious effluent problem. In the near future the water suppliers are more likely to install ion exchange chemical process plant which has, until now, been primarily an industrial process.

The water supply industry's main preoccupation is traditionally with collection and distribution of water. Its technology is controlled by civil engineers (assisted by chemists and microbiologists). It appears that this industry is about to be obliged to move into chemical processing on a massive scale. Ion exchange is not as technologically demanding as the biological processes but its adoption may still have far-reaching consequences.

Ion exchange denitrification

Conventional anion resins take ions in the order

$$SO_4^{--} > NO_3^- > Cl^- > HCO_3^-$$

This means that when a high-nitrate water passes through a column of resin in the Cl^- form it will exchange its SO_4^{--} and NO_3^- for Cl^-. When exhausted, the unit is regenerated with brine (NaCl) and, with counterflow regeneration, the nitrate residual in the product can be made very low. This means that only a part of the flow needs to be treated, and this part can then be blended back with untreated water to give a product which is within specification. Table 18.1 shows the changes in contents.

Table 18.1. Quality of water denitrified by conventional ion exchange resin

Raw water (RW) in blended product %	Water A 10%		Water B 70%		Water C 40%	
	RW	Product	RW	Product	RW	Product
NO_3 mg/l as NO_3	159	45	60	45	81	45
SO_4 mg/l as SO_4	173	43	39	28	58	29
Cl mg/l as Cl	62	226	44	66	34	77
HCO_3 mg/l as $CaCO_3$	197	193	188	181	63	61

Notes: These are three typical nitrate-rich UK waters, denitrified by conventional anion exchange and blended with raw water to a product nitrate level of 45mg/l as NO_3. The product figures assume that variations of product quality during each ion exchange run are eliminated in storage etc. High chloride, especially when accompanied by low bicarbonate, tends to make water corrosive of steel, galvanized steel and brass.

Chromatographic banding and its consequences

The affinity order of these four ions means that, as water passes down the column and as the column becomes exhausted, the ions order themselves into bands in which they are held on the resin. The longer the column, the sharper the separation into these bands. Figure 18.1 shows schematically how they develop.

Assuming that a column is initially in the all-chloride form, the top of the column will first take up sulphate and nitrate in the ratio in which they arrive in the raw water. Bicarbonate is only a little less tightly held on the resin than chloride, but since the resin initially contains no bicarbonate at all, it will also take up some of the bicarbonate in exchange for chloride until it is in equilibrium with the incoming water.

As water continues to flow over the exhausted resin at the top of the column, further exchange takes place in which the resin comes to equilibrium with all the ions in the new water. This means that it will take up more sulphate, which will displace nitrate, chloride and bicarbonate until the resin at the top is almost

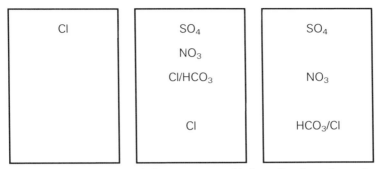

Figure 18.1.   Development of chromatographic banding in exhaustion cycle

wholly in the sulphate form. The water passing on down the column will therefore be richer in nitrate, chloride and bicarbonate than the raw water.

This sorting-out of ions becomes more marked as the ion exchange front goes down the column, and more time elapses in which the ions can separate from one another. Three zones emerge in each of which the resin is especially enriched with sulphate, nitrate and chloride/bicarbonate respectively, as shown in Figure 18.1. This tendency to form bands of ions on the resin has two disagreeable consequences.

- The high concentration of bicarbonate at the head of the ions coming down the column affects the CO_2–HCO_3–CO_3 equilibrium. The high concentration shifts the equilibrium towards the production of H_2O + CO_2. Laboratory trials of the process which are run in glass columns often show a zone in which bubbles of CO_2 are produced, which can cause gas locking in the column. This is not a direct cause of trouble in an industrial scale pressure vessel where the pressure is generally high enough to prevent the CO_2 coming out of the solution.

  When the column is regenerated, the carbonate and bicarbonate are regenerated off the column in such high local concentration that in the presence of hardness there is a danger that calcium carbonate may be precipitated. This precipitation can accumulate and interfere with the process, and an occasional acid wash may be necessary to avoid the build-up causing damage; regeneration with softened water can be used to avoid hard water contact.

- The other problem is that the nitrate concentration in the water passing down the column in front of the sulphate zone is actually higher than in the raw water. If the unit is allowed to continue running past the point at which nitrate breaks through the peak nitrate level in the product can be far higher than in untreated water as the graph in Figure 18.2 shows.

169

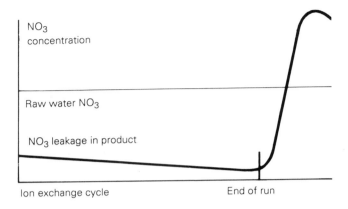

Figure 18.2. Nitrate breakthrough with conventional ion exchange resin

Problems with conventional resins

The straight anion process using conventional resins therefore has several disadvantages:

- As it removes sulphate in preference to nitrate, the effective nitrate capacity on waters with a high sulphate content will be poor; unfortunately we have many waters where this is the case.
- Because the concentration of nitrate in the product coming from the column after breakthrough rises far higher than that in the raw water, the process needs close control.
- All the sulphate and nitrate (and some of the bicarbonate) are converted to chloride. Even when blended back with raw water, this can affect the taste of the water and, more seriously, can make it corrosive. (To alleviate these problems costly bicarbonate has to be used as a supplementary regenerant.)
- The spent regenerant is a strong solution of sodium sulphate and nitrate, together with surplus chloride. Inland disposal of these soluble salts inevitably ends up in the aqueous environment and may be unacceptable on a large scale. Hauling waste solutions for dumping in the sea can become the biggest single operating cost; even this may not be permitted in future.

- The danger of hardness precipitation may have to be avoided by routine acid washing, which is a nuisance and means that acid has to be handled on the site, or by additional plant to provide soft water for regeneration.

Nitrate selective resins

To overcome these problems, the resin manufacturers have developed resins whose selectivity order is

$$NO_3^- > SO_4^{--} > Cl^- > HCO_3^-$$

Inevitably, nature is not cheated so easily and these resins have a lower capacity. They are also more difficult to regenerate than standard resins, the nitrate being difficult to remove and thus requiring advanced regeneration techniques.

However, on waters in which the sulphate:nitrate ratio is high, the nitrate-specific resin will, despite its lower capacity, give a longer nitrate-removing run than the conventional resin, which has to remove both sulphate and nitrate. This means in turn that the effluent which it produces may present a smaller disposal problem than that from the conventional resin. The main benefits from the selective resin, however, are that the chloride level in the product only rises by an amount equivalent to the nitrate removed and, as nitrate is the most tightly held ion, there is no danger of a wave of high nitrate in the product if the unit is accidentally over-run. Selective resins therefore show promise in particular applications: where the sulphate:nitrate ratio in the raw water is high or, on the small scale, where good supervision cannot be guaranteed.

An interesting process uses weakly acid and basic resins, both regenerated with CO_2 under pressure. This results in partial deionization, in which nitrate is reduced, together with all other ions. Its high capital cost can be justified only when the reduction in TDS is generally beneficial.

Other applications

So far we have mainly considered the very large flow rates associated with public supply. In industry, brewers and soft drinks manufacturers, too, are anxious to avoid high nitrate levels in their products, and quite large volumes of water may need to be treated. There are, in addition, cases where water has to be denitrified on a smaller scale.

If high-nitrate water is used in processing canned food, it is liable to attack the tin. Food processors therefore denitrify the modest amounts of water which actually go into the product. The effluent from such denitrification plant is inevitably small compared with the large volumes of waste water from the operation as a whole, so that the disposal problem falls away. Since canned foods contain added salt, the small additional amount of chloride from the denitrified water is harmless. Ion exchange denitrification using conventional resins has already given useful service in such applications for some years.

Membrane processes, reverse osmosis (RO) and electrodialysis, also show potential for nitrate reduction, especially where general reduction of TDS is desirable. One large RO plant is now (1990) giving excellent service at a brewery in the north of England.

There are a large number of small private water supplies which require denitrification. These typically supply isolated farms and, surrounded as they are by the very activity which puts the nitrate in the water, some of these supplies exhibit very high nitrate levels. Here the nitrate is best removed from a separate pipe reserved for drinking water, leaving the bulk of the supply untreated for washing, toilets and so on. Nitrate-specific resins are normally best suited for this, and the resin can be installed in a modified water softener.

For the very small scale, there are now proprietary devices containing a small non-regenerable charge of resin; their cost per litre of water produced is very high, but the capital investment is low and they avoid the inconvenience of regenerating the exchanger.

Costs

Water costs to the consumer vary enormously. In public supply, the main component in the price is the cost of distribution; the 'ex works' cost may be about $5p/m^3$. Large-scale denitrification may add as much again, but in a 'delivered' price of $30p/m^3$ this can perhaps be tolerated.

Bottled mineral water costs 1000 times more. At 57p/litre for a well-known brand, it is actually dearer than petrol. The cost per litre of water treated by non-regenerable devices is hard to assess but still far cheaper than bottled water.

Development of ion exchange processes

Various improved processes are being explored but cannot yet be considered for large installations. Full-scale plants for denitrification to be built in the near future are likely to be conventional in operation and filled with conventional or nitrate-specific resins, depending on the conditions at the site.

One possibility is to combine the biological and ion exchange processes. The water is treated by ion exchange and the spent regenerant is subjected to biological denitrification, after which it can be recycled and used again in a subsequent regeneration. In breaking down the nitrate, the bacteria create an almost equivalent amount of bicarbonate; the recycling regenerant solution thus becomes richer and richer in bicarbonate. The ion exchanger can be regenerated equally well with bicarbonate and the process thus requires only limited supplementary chloride to maintain the total quantity of regenerating ions in the solution. However, sulphate builds up in the recycling regenerant. The problem of sanitizing the resin after regeneration with biomass appears to make the process impractical.

Another process uses an intermediate stage in which the resin is put into the SO_4^{--} form using sulphate recovered in the process. It promises reduced regenerant usage and efficient production.

Summary

Ion exchange is now being used for denitrifying water for public supply as this becomes increasingly necessary. It is at present both cheaper and safer than the alternatives.

In the long term, this situation may change as intensive worldwide efforts are made to find alternatives or improvements. The main disadvantage of chloride ion exchange is a tendency to increase the corrosiveness of the water which may in turn require post-treatment to correct, and research into ways of reducing this problem continues.

The medical aspects of nitrates

Nitrate is a water-soluble ion which occurs naturally in the environment as a consequence of nitrogen fixation from the vast reservoir in the atmosphere. The scale is substantial. There is, for instance, estimated to be 10 000kg of organic nitrogen per hectare of average British soil.

Plants take up the various nitrogen compounds from the soil, and they thus become available to animals. They are also present in some water supplies but usually at low level. Nitrate is therefore almost certain to be in everything that we eat and drink, and humans have been exposed to this ion throughout their evolution. However, a comprehensive report on the subject, 'Nitrate, Nitrite and N-Nitroso Compounds in Food', was published by HMSO in 1987 for the Ministry of Agriculture and Fisheries. This quotes the Department of the Environment as having shown in 1985 that 103 public water supply areas (representing 11 surface water and 92 ground water supply regions, and serving about 2 million people) had levels of between 50 and 100mg/litre of nitrates at some time in the previous two years.

Any threat to our health comes from the fact that nitrate can be converted in humans to the nitrite ion. Nitrite can react with the oxygen-carrying substance haemoglobin in the blood, causing it to be less efficient in its difficult and responsible task. The condition is referred to as methaemoglobinaemia, and it is far more likely to be serious in artificially-fed infants under 3 months of age than in adults. In this respect, the nitrate content in the liquid of a bottle feed is liable to be much more dangerous than the same material in solid nourishment.

Nitrate can also combine with secondary or tertiary amines to compose N-nitroso derivatives, a reaction known as nitrosation. Some N-nitroso compounds have been demonstrated to provoke cancers in a wide range of laboratory animals.

In his book *Cell Tissue and Disease,* Professor Neville Woolf (Bland Sutton Professor of Histopathology at the Middlesex Hospital Medical School) mentions that the oncogenic action of the nitrosamines and nitrosamides was discovered by Magee and Barnes in 1956. They were carrying out toxicity studies with dimethylnitrosamine in the rat. Nearly all the animals developed malignant tumours in the liver after 26 to 40 weeks at a level of 50 parts per million in their diet. Since then, he notes, the nitrosamides have also been shown to be toxic, mutagenic and teratogenic.

Epidemiological data vary. In Britain there is apparently no detectable ill effect in adult consumers of high nitrate waters. Studies which do report such effects suggest that they are linked to poor hygiene and/or poor nutrition.

Part 7
MEMBRANE PROCESSES

19   Membrane processes in general and electrodialysis in particular

All the processes which have been described in previous chapters have been in common use for decades, though of course there has been continuous progress, especially in the detail of plant designs. Membrane processes, as such, are also very old but their introduction into water treatment is much more recent and they are, therefore, relatively unknown quantities. At present, they represent a means of carrying out special duties, usually at a greater cost than the conventional processes. Their costs, however, are falling all the time, and at the same time the special requirements for which they are necessary are becoming commoner. We can therefore expect to have more and more to do with membrane processes in future.

What is a membrane

For our purposes all membranes are thin films of solid materials which have some selective property, that is, they allow some classes of substances to pass through them but not others. Membrane processes which make use of this selectivity share a number of basic features,though there are several different kinds of membrane and membrane process.

The basic features of all membrane processes

To use the properties of a selective membrane we need an apparatus which has the following features:

- it has to have two sets of liquid conduits, one for the water we are going to treat, and another for treated water;
- it has to be built so that the membrane separates these two conduits in such a way that there is no direct cross-leakage between them;
- it has to provide some kind of driving force which will force through the membrane the substance which the membrane allows to pass.

In addition to this, the whole group of processes has intrinsic chemical enginering features which are the subject of this chapter. These are bound up with the theory of mass transfer.

Three membrane processes

Three processes have come into regular use in water treatment, based on three different kinds of membrane:

- The reverse osmosis (RO) membrane comes in very different forms, all with the same essential property: each allows water and very small dissolved, undissociated molecules to pass through, but will not pass ionized salts, large molecules, or colloidal and suspended solids. The driving force which pushes the water through is hydraulic pressure.
- The ultrafiltration (UF) membrane is really an extremely fine filter which can be made with pores of different desired sizes. It allows water and small dissolved materials to pass through it, but retains larger molecules, colloids and suspended material. The pore size determines the minimum size of molecule which will pass. It, too, uses hydraulic pressure to push the water through.
- Electrodialysis (ED) depends on two different membranes, made essentially of cation and anion exchange material. These two membranes allow small cations and small anions to pass through respectively, but they retain everything else, that is, water, large ions of both signs, non-ionic dissolved materials, suspended and colloidal material. The driving force for ions in electrodialysis is an applied electromotive force.

Mass transfer

Mass transfer is one of the basic subjects studied by chemical engineers, and the simple explanation which follows is intended for those reared in other disciplines.

The basic features of mass transfer are similar to those of heat transfer (which will be familiar to mechanical engineers). In membrane processes we have to transfer some material from one liquid phase to another, across some kind of separation barrier. This means transferring salts (in ED) or water (in RO and UF) from one solution to another.

Three variables immediately present themselves as the factors controlling the rate of transfer:

- the resistance to transfer per unit area of barrier;
- the driving force promoting transfer; and
- the area across which transfer takes place.

(This is, of course, the same pattern as in electrical power transmission where these three variables are represented by the resistivity of the conductor, the voltage and the area of the conductor, respectively).

As in heat transfer, the resistance of the separation barrier (in our case the membrane) is composed only in part of the resistance of the barrier itself. The material which we are trying to transfer across has to get to the surface of the barrier on the one face and away from the surface on the other face.

As the barrier presents a discontinuity to flow, there is always a stagnant layer of fluid on its faces. The thickness of this stagnant layer varies with the hydrodynamics of the system, conventionally characterized by the Reynolds number (see below).

Suppose we are looking at an ED anion membrane and we are trying to get chloride ions to go across the membrane under the influence of the electrical field. Figure 19.1a shows the conventional picture by which the process is usually illustrated. The membrane is represented as being flanked by two completely stagnant films of water; outside these stagnant films, the water is supposed by in a state of total mixing. In fact this is a slight over-simplification but the convention serves chemical engineers very well.

In the turbulent region, where there is perfect mixing, the concentration of chloride ions must be constant, because any change in any part of this region will immediately be evened out by the turbulence. On the other hand, the stagnant water films present an obstacle to diffusion of chloride ions. It so happens that the films may present more of an obstacle to flow than the anion selective membrane itself. It is therefore rather important to design the system and its hydrodynamics in such a way as to maximize the turbulence measured by the Reynolds number, in order to decrease the film thickness and so reduce the resistance to mass transfer. This sort of argument will be familiar to all those who have studied heat transfer, where the copper tube presents less of an obstacle than the water films on it.

The Reynolds number, N_{Re}, which governs the turbulence and hence the film thickness is a dimensionless number $N_{Re} = vl\rho/\mu$.

179

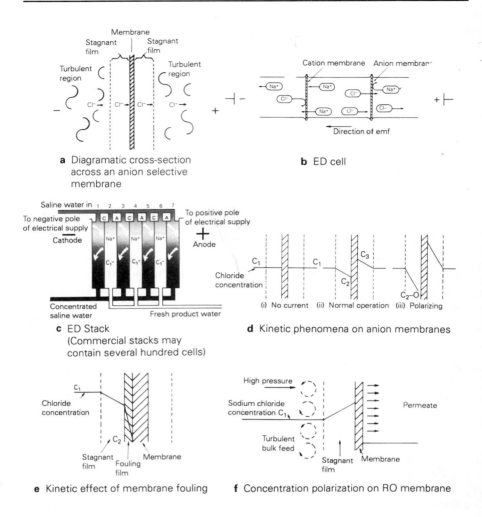

a Diagramatic cross-section across an anion selective membrane

b ED cell

c ED Stack (Commercial stacks may contain several hundred cells)

d Kinetic phenomena on anion membranes

e Kinetic effect of membrane fouling

f Concentration polarization on RO membrane

Figure 19.1. Electrodialysis and reverse osmosis membranes and concentration phenomena

where v is the velocity of flow, l is a linear dimension of the gap through which the fluid is flowing; for example in pipes it is the diameter, and in flow between parallel plates it is the distance between the plates: r is the density of the fluid and m its viscosity. In mass transfer (as in heat transfer) the designer will therefore attack the variables in the Reynolds number with a view to increasing N_{Re}. He will, for example, increase v as far as he can without incurring excessive pumping costs. He will try to increase l but usually there are important reasons for keeping l small, so here again he will compromise. r is generally outside his control but m, the viscosity, falls as the temperature rises. All membrane processes work better at higher temperatures, subject again to limitations imposed, for example, by the material of the membrane.

Again, as in heat transfer, any additional film which obstructs the transfer will add an additional resistance to mass transfer. Such films can easily become the controlling element in the process.

Polarization ※

The situation becomes more complicated when we consider that the basis of all membrane processes is the selectivity of the membrane used. The RO membrane, for example, is bombarded by molecules of water and ions of the salts dissolved in the water. The process is based on the ability of the membrane to allow water to pass through while rejecting the ions.

Imagine the concourse of a railway station. Passengers are milling about and want to go through the turnstile to the train but the ticket collector rejects those without reservations. These now have to find their way back into the middle of the concourse, against the stream of more passengers converging on the turnstile. If the rush to get to the train is very great, we could end up with the turnstile completely barricaded by hapless, reservation-less passengers trying to get back, thus preventing the movement of passengers to the turnstile at all. Even if this extreme state of affairs never arises, the proportion of passengers without reservations will be higher in the crowd at and near the turnstile than in the concourse in general.

If, however, the whole station crowd were kept much more on the move, then the rejected travellers would be dispersed more readily and the congestion at the turnstile reduced correspondingly. Exactly this state of affairs occurs on the surface of an RO membrane, where the rejected salt ions will be found in much greater concentration than in the bulk of the liquid.

To summarize:

- All membrane processes depend on mass transport across a selective membrane.
- The rate at which this transport takes place depends on the resistivity of the separating barrier per unit area. The separating barrier in this context includes the liquid films which inevitably form on the membrane surfaces and any fouling film which may accumulate in addition.
- The amount of transport also depends on the total area of barrier available for transport and on the driving force available.

- Because of the selectivity of the barrier, concentration changes occur in the stagnant liquid films on the membrane surfaces. The degree to which the concentration changes occur depends on the rate of transfer and on the thickness of the film.
- If the concentration change is allowed to become very great, it will eventually reach some limiting condition where it has some major and unforeseen effect on the transfer process.

This concentration phenomenon in the stagnant film on the membrane surface is called polarization. It has proved one of the most important and most intractable features of all membrane processes. Its actual effect, however, is different in each different process.

Electrodialysis and polarization

Anions = neg ions
Cations = positive ion

Suppose we fabricate anion exchange material in the form of a thin sheet (Figure 19.1a). We know it is a porous substance with fixed positively charged groups on it; these groups must have negative ions associated with them, but the negative ions can move in and out of the materials, always provided there are other ions available to replace them. Ion exchange processes use the preference which these materials have for some ions rather than others, but in ion exchange membrane processes, these preferences can be ignored as a first approximation.

If we dip our anion exchange sheet in a bath of salt solution and apply an EMF across it, then there is nothing to stop chloride ions from moving towards the positive electrode, either in the bath or across the membrane. An electrical current will be set up; in the solution it will be carried equally by sodium and by chloride ions moving in opposite directions but in the membrane it will be carried by chloride ions alone, because an anion exchange material does not admit cations into its pores. We now have mass transfer of chloride across the anion exchange membrane (called the anion membrane for short). The same thing can be done with a sheet of cation membrane, which will transport only sodium ions.

Now suppose we make up a cell as shown on Figure 19.1b with a cation and an anion membrane side by side. If we pass a current across the apparatus, sodium and chloride ions flow across the membrane and out of the compartment formed by the two membranes, so that the solution between the membranes becomes desalted. If we go a stage further and make up an apparatus in which cation and anion membranes alternate, as shown in Figure 19.1c then it will be seen that the current will desalt all the even numbered compartments 2,4,6 and so on, while the odd-numbered compartments become enriched in salt, with sodium and chloride ions streaming in from the

182

two membranes which form the compartment. We can, therefore, build an apparatus in which cation and anion membranes alternate, and if the cells formed between the membranes are connected to two liquid manifolds so that all the even compartments flow into one and the odd compartments flow into the other, then we have a means of desalting a stream of water (or of enriching the other one). This is called an electrodialysis stack (ED stack for short).

In ED, electrical currents flow through the water being treated. As pure water is almost an insulator, and water really only becomes conductive by virtue of the dissolved salts in it, this process is no good for totally demineralizing water. Its main use in water treatment is for desalting brackish water to potable level. (The enrichment of the concentrating stream has also been used by the Japanese for concentrating sea water into strong brine.)

Even in brackish water conversion, ED has proved quite troublesome in some cases. Its main use is in industrial separations rather than water treatment, so there is no need to describe ED equipment in detail. But the problem of polarization in ED is rather an interesting example of the kind of difficulties which we experience with membrane processes. Let us look at the mass transfer across anion selective ED membranes.

Figure 19.1d(i) shows the membrane with a salt solution on both sides of it before the current is switched on. We have the usual stagnant films on the membrane face but as there is no mass transfer taking place yet, the concentration of chloride in the films is the same as in the turbulent bulk of the liquid and is, say, C_1. If we apply an EMF across the solution and membrane, then the chloride ions will be driven across from left to right, as shown on the diagram.

The membrane itself is an excellent conductor, and there is a rush of chloride ions across it, with the result that the left-hand membrane surface begins to gulp chloride ions from the liquid touching it, while the right-hand surface pumps chloride ions into the water. The concentrations on the membrane face are therefore changed. The left-hand face becomes depleted, and so a concentration difference between the membrane face and the bulk of the liquid arises. Under the influence of this difference, chloride ions diffuse across the stagnant film, forming a concentration gradient. On the right-hand face, the opposite happens, with a high concentration on the membrane face and a concentration gradient away from the membrane.

After an instant, electrical currents across each of the two films and across the membrane have to become the same, because they are electrically in series. The concentration gradients therefore end up such that the flow of ions in all parts of the apparatus is the same, and this is the condition shown in Figure 19.1d(ii). Here the concentration in the solutions is C_1, but on the depleted left-hand face of the membrane it has fallen to C_2 and on the enriched right-hand face of the membrane it has risen to C_3.

If we increase the EMF, the current flows faster and the gradients have to become steeper. This is all very well but there is a theoretical limit to the amount of current we can push through the apparatus. Consider what happens in the depleted film on the left-hand side of the membrane. As we raise the EMF, and the current flows faster, we have to have a steeper gradient. As C_1 is constant, and assuming a constant thickness of film, this can only come from a progressive lowering of the chloride concentration on the membrane face, C_2. Eventually we reach a point at which C_2 falls to zero, that is, we are exploiting the maximum concentration gradient available to us (Figure 19.1d (iii)). Any further increase in EMF will then produce no further increase in current. 'Depletion polarization' has set in and is accompanied by the most undesirable side-effects, which we need not go into. The way to increase the rate of transfer would be to reduce the film thickness, or raise the temperature, to reduce the resistance of the film.

There is worse to come. Suppose our saline water contains large organic anions such as the humic and fulvic acids described earlier (see p.12). Under the influence of the EMF they, too, migrate to the membrane but they are too large to get into the pores, let alone pass through the membrane. They settle on the surface of the membrane where they form a film which has a high resistance to diffusion by chloride ions (Figure 19.1e). This film acts like scale in a boiler, and becomes the major resistance to mass transfer. Juggling about with the hydrodynamics to make the water films thinner will now do very little to improve the process.

Polarization in reverse osmosis

The RO membrane lets water pass through and practically nothing else. Figure 19.1f shows the section across the RO membrane. If we start with a concentration of C_1, and apply the pressure which squeezes the water through the membrane, we can see what is bound to happen. As water begins to go through the membrane the salts in it get left behind on the membrane surface, so their concentration rises. A concentration gradient is then set up which promotes diffusion of these salts back into the bulk of the fluid, across the film. Eventually an equilibrium is set up, when the diffusion rate back across the film is the same as the rate at which water passing through the membrane leaves salt behind on its surface. If there are colloids or suspended materials in the water as well as salts, they will simply settle on the membrane surface and stay there, forming an additional barrier to diffusion. Both the local rise in concentration and the additional barrier are important to RO operation.

Polarization in ultrafiltration

The problem is similar to RO, but is not technically called polarization. The UF membrane does not retain salts but only large molecules and bigger particles. The problem of concentration build-up of salts does not arise, but otherwise the operation is similar to RO. Large molecules diffuse much less readily than salt ions and somehow or other they have to be got off the membrane face if the process is to run continuously. Colloids and bigger solids are even worse. We shall go into the detail in Chapter 21.

✗Electrodialysis

We can now say a last few words on electrodialysis. As we have seen, it only operates economically where there are enough dissolved salts in the water to conduct electricity. This limits the process to brackish water conversion. Its ideal economic range is for desalting water of less than 2000ppm down to 500ppm. If we need a lower product TDS, ion exchange has to take over because ED becomes uneconomical. If we start from a high TDS, RO can do the job more economically. ED cannot deal with water containing organic or colloidal materials, except after pretreatment, because very severe filming of the membranes takes place as described above. Where impurities of this kind have to be removed, RO can do it (though there may be fouling problems). Altogether, ED in its present state of development is limited to a narrow range of applications in water treatment, and can be ignored by all but a few specialists.

An old variant of ED has recently re-appeared as a commercial process: the cells of the ED stack are filled with ion exchange resins to provide a path for ions. This allows the process to be used for producing very pure water without using chemicals for regeneration.

20 Reverse osmosis

The three membrane processes mentioned in Chapter 19 were none of them intended originally for purifying water in temperate zones; ED and RO started out as methods of converting brackish water to drinking water, and UF was aimed at recovery or purification of process liquids. Generally speaking, the three processes have found their main use in these original aims, but it turns out that they also have uses in the field of high purity water production.

RO has been the most successful of the three processes in general, and has also found the widest use in water purification. It performs two quite different duties, for either of which we might consider its use:

- RO will remove from 90–98 per cent of all dissolved salts in a raw water, the removal depending largely on the type of membrane used.
- RO also removes large dissolved organic molecules, all colloids and suspended matter, and micro-organisms.

In removing salts, RO does the same job as ion exchange but with a less pure product. With ordinary potable water quality as starting-point, RO is rarely economical for industrial use and ion exchange is normally used for demineralization. At more than 400ppm TDS, however, it is worth considering RO as a first stage on purely economic grounds, and the higher the initial TDS, the better the prospects for RO become.

On the other hand, the power of RO to remove all large molecules and all particles makes it a useful stage in the production of ultra-pure water. It is frequently used, for example, in the micro-electronics industry to remove organic and colloidal material from raw waters.

One major UK electric component maker has a works which draws on such a clean water, derived from the chalk measures around the Marlborough Downs that, with ion exchange alone, he can obtain 18 megohm water, that is, of theoretical conductivity. Most of Britain's other electronics firms have set up where they use some of the most difficult waters in Britain — in Scotland, in Wales and on the Bedfordshire Ouse. They would be unable to operate without RO to help produce the ultra-pure water needed for washing microchips.

The basis of the process

In simplified terms, RO consists of pushing water through a membrane under hydraulic pressure. The hydraulic pressure has first to overcome the osmotic pressure caused by the difference in salt concentration on the two sides of the membrane, and the surplus pressure then supplies the driving force pushing water through the membrane. When treating low TDS waters for our particular purpose, the osmotic pressure can largely be ignored.

Process variables

As a first approximation we can say that the flux of water across the membrane (that is, the flow-rate per unit area) varies with the applied pressure divided by the resistivity of the membrane itself. We therefore get an equation which is identical to the one which describes the flow of electricity through a conductor:

$$\text{Flow} = \frac{\text{pressure} \times \text{area}}{\text{resistivity}}$$

Just as in electrical engineering, where we have the choice of voltage, cable diameter and cable material, in RO we have these three variables at our disposal.

The pressure used for RO in water purification applications is usually about 10–40 bars although, in some very small plants where the capital cost of the membrane is small, lower pressures are used for convenience and so under-utilize the membrane. Sea-water conversion needs higher pressure.

The really interesting variables are those of area and resistivity, because two completely different membrane systems have been developed:

- the sheet membrane (originally cellulose acetate, which is still widely

187

used) which is expensive per unit area but has a low resistivity; and
- the hollow fibre membrane, which consists of polyamide capillaries the thickness of human hair, which is very cheap per unit area but has a high resistivity.

The sheet membrane is most commonly used in spiral wound modules in which it is wound round like a Swiss Roll. The raw water flows along the axis of the roll, under pressure. The permeate which flows across the membrane flows spirally out through a separate outlet (see Figure 20.1).

Plate 14.   Typical pumping installation for membrane plant

The hollow fibre module consists of a U-shaped hank of fibres enclosed in a pressure vessel. The open ends of the U are sealed by being cast into an epoxy tube plate (see Figure 20.2) rather like a floating head heat exchanger.

We shall see how these two different approaches result in RO being used in rather different ways.

Plate 15. Unusual photograph of a large reverse osmosis plant in an oil refinery

Product purity

It is not yet clear just how the RO membrane works. One odd fact is that the salt passage (the ratio of permeate TDS to feed TDS) of a membrane is constant over a wide range of conditions. Since higher pressure leads to higher flux this constant salt passage results in lower percentage salt residual in the permeate, another reason for using high pressure. As membranes age, their resistivity rises and the flux drops, and so does the treated water quality. Salt passage tends to increase with age which makes the situation worse.

As might be expected, the RO membrane rejects large multivalent ions most efficiently, so that Ca^{++}, Mg^{++} and SO_4^{--} are taken out to a greater extent than Na^+ and Cl^-. In some membranes this selectivity has been enhanced so that they remove most of the hardness from raw water at relatively low pressure, and may be used for partial softening. Silica passage varies, and up to half the silica can be passed through, depending on circumstances. The membrane does not reject CO_2 at all. Small non-dissociated organic materials dissolved in water pass through unchanged, and so do dissolved gases like O_2.

The stagnant film

We have described the stagnant film on the membrane surface which occurs in all membrane processes, and has a major influence on the operation of each of them. The first important factor is that the membrane only sees the solution as it exists inside the film, in the membrane face. As the film on the raw water side in RO becomes more concentrated than the bulk solution, the percentage of salts passing through the membrane is based on this more concentrated solution and not on the bulk of the solution, away from the membrane. The concentration in the stagnant film must, therefore, not be allowed to rise too high, or the product will suffer.

Another danger, of course, is that if the concentration in the film gets beyond the solubility of some of the salts in it ($CaSO_4$ is the most obvious danger) then scale will form on the raw water side of the membrane due to the precipitation of the super-saturated salts.

We have shown that the concentration changes which take place in the stagnant film depend on two factors: the degree of turbulence in the water, which controls the film thickness, and the rate at which the membrane process is functioning and at which (in the case of RO) it piles up salts on the membrane face. The two alternative membrane systems handle the polarization problem in different ways.

The flux across a spiral wound membrane is high and salts accumulate quickly at the membrane surface. High concentrate flows are necessary to sweep away these salts. The configuration allows for these high flows, and

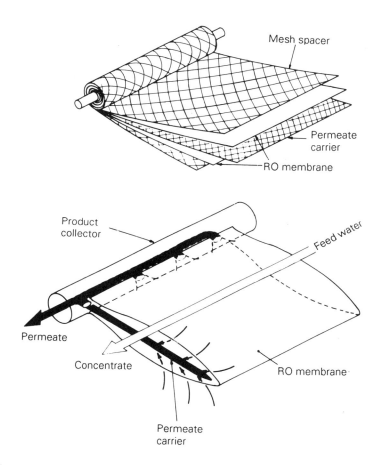

Figure 20.1.   Spiral wound module

spacers in the Swiss roll are designed to promote turbulence.

By comparison, fluxes in hollow fibre membranes are an order of magnitude lower. Salts therefore accumulate much more slowly at the membrane surface. The configuration of a hollow fibre module is such that concentrate flows across the membrane surfaces are inherently lower. Since the salts accumulate slowly this is acceptable.

Conversion

The difference in concentrate flow rates is important because it affects

191

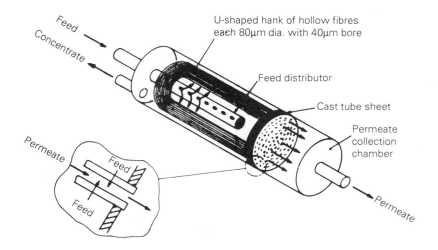

Figure 20.2. Hollow fibre module

conversion, that is the permeate flow expressed as a percentage of feed flow. The configuration of a hollow fibre module with its low concentrate flow rate can give up to 55 per cent conversion. A single element in a spiral wound system rarely achieves more than 10 per cent and to achieve 50 per cent in such a system requires six elements in series, resulting in a module considerably longer and usually more expensive.

In large plants, this disadvantage can readily be overcome by an arrangement of modules in series, and parallel. Concentrate from the first stage is used as feed to the second and so on, each successive stage employing fewer modules as the flow reduces. Figure 20.3 shows a typical arrangement. In small plants, where there are not enough modules to make this kind of arrangement possible, the spirally wound membrane remains at a disadvantage.

On the face of it, a low conversion does not seem to be a serious problem; water is cheap (or so a lot of people think) and anyway the waste from an RO plant producing ultra-pure water can be re-used for cooling or for rough washing.

In fact a low conversion is likely to be rather costly. The water fed to an RO plant has almost inevitably been pretreated to remove dirt, to adjust its pH and to remove its hardness, or dosed to avoid scaling. These costs are referred to the flow of raw water pumped in and, together with the cost of pumping, form the main running costs.

Moreover, a low conversion means that the pretreatment equipment and the

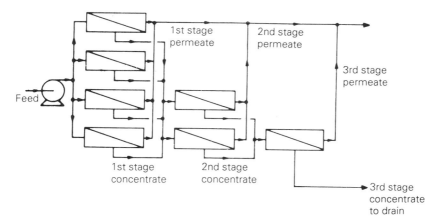

Figure 20.3. **Three-stage RO system with concentrate recovery**

high-pressure pump all have to be oversized, and so the capital cost will also be affected.

The choice of module therefore becomes straightforward in most cases. Really small plants cannot function economically with spirally wound modules, because of the recovery problem. Below, say $10m^3/hr$, the hollow fibre module is likely to be the economic choice, although other considerations can apply.

Physical blocking

Particulate and colloidal material, whether organic or mineral, will not pass through RO membranes but may cause physical blocking. The higher surface velocities in spiral wound modules make them less prone to this type of blocking and easier to clean than the hollow fibre type. This is a particularly important factor in the choice of larger plant.

The ability of a particular water to foul a membrane is characterized by the silt density index (SDI). This is a measure of the rate of build-up of resistance to flow and is determined by filtering the water through a 0.45μ membrane filter and measuring the time taken (t_0 minutes) for 500ml to pass through at 2 bar inlet pressure. Flow through the filter is continued for a further period, T minutes, at the same pressure and the 500ml test is repeated giving a new time (t minutes).

The SDI is given by

$$\text{SDI} = 100 \left(\frac{t - t_0}{t}\right)/T$$

193

That is, the percentage change in flow rate divided by the time taken for the change to occur.

Another statement of fouling potential, and one that is not specific to the particular test period, T, is the plugging factor, which is simply the SDI multiplied by the duration of the test in minutes, that is $PF = T \times SDI$. This is a derivative of the test and it is normal to state the pressure at which the test is carried out, for example PF_{30}.

There are other versions of these formulae, and most membrane manufacturers will provide advice as to how their particular index is measured and what index their membranes will tolerate, so that the need for pretreatment can be assessed. A good rule of thumb is that the SDI (as defined above) of RO feed should be less than 3 and the feed should be filtered to 5μ or less.

Some more variables

The two systems, with the conventional type of membrane fitted, vary in other respects. Polyamide is extremely sensitive to oxidation by chlorine, which must be removed from the water. Cellulose acetate, whilst still suffering some deterioration, will tolerate up to 1ppm of Cl_2 in comparative safety.

Both types of membrane suffer degradation by hydrolysis and this is accelerated as the pH of the feed deviates from neutral. Cellulose acetate membranes deteriorate rapidly if the pH falls outside the 4–6 band. Polyamide hollow fibres are more tolerant of high pH and can operate in the range 4–11.

A new generation of membranes is currently overcoming these drawbacks. It is made by depositing an extremely thin film of actual membrane on a porous substrate which is made of another chemical, hence the name thin film composite (TFC). Further rapid development in membrane technology can be expected.

The resistivity of the membrane system depends on the viscosity of water, and so falls quite sharply if the water is warmed. If the product is going to be heated anyway, then large economies can be made by preheating the raw water and reducing the size of the RO plant for a given output. The spirally wound membrane can withstand up to about 30°C, and the hollow fibre module about 35°C but, of course, the rate of hydrolysis is increased at these higher temperatures.

Bugs and ultra-pure water

One of the problems in producing ultra-pure water is keeping it free of microbiological growth. RO does remove bacteria (and even viruses) efficiently, but that is no guarantee that they will not breed again in the purified water. If they grow on the RO membrane itself, of course, they may block it. The hollow fibre membrane, in particular, is a very good breeding-ground for bacteria, especially as it cannot be kept sterile with chlorine.

In theory, no bacteria should ever get through to the clean side of the membrane, but special precautions should be taken against contamination creeping up the treated water lines, which should, for example, be isolated from atmospheric contact. To inhibit bug growth when using RO in these circumstances, routine sanitization of the modules is advisable.

Flow sheets for ultra-pure water

The main users of ultra-pure water are the electronic component manufacturers, whose main preoccupation is to have water of the lowest possible conductivity, and the pharmaceutical users who want to exclude bacteria and pyrogens and are less interested in total removal of inorganic salts.

These two objectives lead to two different types of flow sheet. The electronic users install RO ahead of ion exchange but after whatever pretreatment is necessary to protect the RO plant from fouling. In this way they get the benefit of partial demineralization before the ion exchange plant removes the last traces of salts to give a product of 18 megohm.

Since its acceptance by the US pharmacopoeia for producing pyrogen-free water, the use of RO has grown in the pharmaceutical industry when this quality of water is specified. Here it is used as a final polishing stage after deionization has reduced the mineral content to acceptable limits.

Summary

RO is (for the present, at least) a relatively expensive process for demineralization and, given conventional sources of potable water, it is not normally economical merely for that purpose.

Its main use in water purification relies on its ability to remove a large group of materials which no other process can be guaranteed to remove. It is therefore widely used when ultra-pure water has to be produced.

This picture may change in the future. As our water supplies get worse, and RO technology improves, we may find an increasing number of applications for RO as a first-stage demineralization process or for rough softening.

21 Ultrafiltration and pretreatment for membrane processes

Ultrafiltration is the third membrane process used in water treatment, and the most recent. It is at present most useful for the production of ultra-pure water though it is not yet clear to what extent the process will establish itself in practice.

What's in a name? Just when RO as an abbreviation for reverse osmosis has become universal, we start to come across people who prefer to speak of 'hyperfiltration'. What is the point of this cumbersome new word?

The best clue here is that the manufacturing company which prefers hyperfiltration to RO makes a whole series of membranes, with a wide range of pore sizes. The smallest-pore membranes are intended for what we will obstinately continue to call RO; they reject dissociated ions by a process which, as we have seen, is still not perfectly understood, and also filter out all molecules larger than about molecular weight 300. At the same time they hold back all colloids and suspended particles.

With a larger pore size than these RO membranes, the membrane's power to reject ions rapidly disappears but it can still serve as a filter, capable of straining out molecules and particles of progressively larger sizes. There is no mystery about the mechanics of this process, which is simple tea-straining with a very fine sieve.

Presumably the process by which large molecules and particles are rejected from an RO membrane is also simple tea-straining. With respect to that kind of use to which RO can be put, it is therefore reasonable to call the process hyperfiltration. Moreover, RO applications merge into ultrafiltration without any very clearly defined distinction, so there is something to be said for having a similar name for the two processes.

Ultrafiltration

We thus apply ultrafiltration to a true filtration process designed to deal with the smallest filterable particles down to molecules of, say, molecular weight 5000, but without the accompanying reduction in dissociated salts which we get in RO.

RO membranes can be made loose or tight to control their flux and rejection properties. UF membranes are made by an extension of these techniques but with a looser structure so that the salt rejection is zero. UF membrane manufacturers have learned to control the pore size with some accuracy and uniformity, and make membranes designed to have cut-off properties for different sizes of molecules.

The membranes are asymmetrical; like RO membranes, they consist of a thin skin which is the actual membrane, which sits on a porous substrate whose duty is to give strength and act as a carrier. This thin-skin technique means that the pressure loss across the membrane is kept to a minimum. As there is no need in UF to choose a membrane material which will have semi-permeable properties, UF membranes can be made in a much wider range of materials and therefore offer a much wider range of resistance to aggressive chemical conditions.

Further up the range of pore sizes for UF membranes we find ourselves merging into membrane filtration, a much older process. Classical membrane filters are developments of the familiar chemist's filter paper: they are mats or sinters, of controlled uniform pore size through the whole depth of the membranes. Modern membranes for membrane filtration are, not surprisingly, beginning to look like UF membranes. The distinction is therefore an artificial one. Quite arbitrarily, we could say that the lower limit of membrane filtration is 0.2 microns; anything with a smaller pore size is ultrafiltration.

The accumulation of large particles on the surface of a UF membrane effectively creates another membrane whose pore size tends to be smaller, and which not only adds to the pressure loss but rejects particles smaller than the membrane cut-off suggests.

UF apparatus

We have seen that in all membrane processes one of the controlling problems is how to disperse the concentration changes which take place on the working face of the membrane. In ED the working face is depleted, and in RO the problem is due to rising concentration of salts on the membrane. In UF we are faced with a similar problem: the molecules and particles which have been retained on the membrane face have somehow to be got back into the bulk of the concentrate and their accumulation on the membrane face has to be stabilized at a workable level.

In UF we are rejecting only large and therefore sluggish particles from the membrane face. These diffuse slowly and would like nothing better than to settle down on the membrane surface. Getting them away from there is the main limitation in UF operations.

On the other hand UF membranes, with their large pores, offer a low pressure loss provided they can be kept clean. In the absence of any salt rejection, there is, of course, no question of osmotic pressure opposing the applied pressure as happens in RO. UF plants thus operate at much lower pressures than RO.

We have previously seen (see pp.190–92) that there are two ways of dealing with the concentration effect on the membrane surface: one is to limit the flux in order to control the rate at which rejected material builds up, and therefore the rate at which it has to be dispersed. The other is to create as much turbulence in the liquid as possible and so reduce the thickness of the film across which the accumulated material has to diffuse. The same arguments apply to UF, only even more strongly, because of the sluggish nature of the materials with which we are dealing. In UF apparatus and design, therefore, it is important to obtain the highest possible liquid turbulence on the concentrate side. At the same time it is necessary to keep the flux low, even though the low hydraulic resistance of the membrane itself seems to offer the prospect of high fluxes.

If the character of the application is such that the membrane is removing little or no material, then the flux can be quite high. With a heavy load, on the other hand, the normal flux on UF is not greater than it is with RO; but of course this flux is obtained at far lower pressures.

As far as configuration is concerned, the fine capillary membrane is out of the question because it is impossible in practice to get high turbulence in a vessel packed with fine fibres. The smallest practical tubular UF membranes are about 0.5–1.5mm in diameter. In contrast to RO, these are used with the pressure and the working-surface on the inside of the tube. The advantage of small tubes is that they are made with their own integral reinforcement in the form of a specially powerful substrate, and then they are strong enough to resist both internal and external pressure without further support. Larger tubes are also used, of the order of 10mm diameter, but they need to be carried in a porous supporting tube; the apparatus physically resembles a heat exchanger in its layout. Yet another manufacturer uses flat sheet membrane carried on moulded plastic plates assembled into a stack whose construction resembles that of a filter press.

Applications for UF

UF is the youngest of the three membrane processes, and we are still finding out what kind of duties it performs best. Its earliest uses have, naturally enough, been to recover valuable materials, when the lure of large profits makes risk-taking most attractive. As a result, UF is in widespread use where a suspension, or emulsion or a solution of large molecular weight materials has to be concentrated, and preferably when at the same time unwanted dissolved or small-molecule impurities can be washed away.

A good illustration of what UF can do in this way (though by no means the commonest application) is the production of de-alcoholized beer. Ordinary beer is recirculated through a UF plant so that water and alcohol pass through the membranes whose pore size is chosen such that they retain all the large molecules which give the beer its taste and aroma. The water loss is made up by adding fresh water, so that the only change in the beer is a progressive lowering of the alcohol content, until the resulting fluid (it is hard to bring oneself to call it beer) is fit to be sold, for example, at motorway service stations. Alternatively, a beer concentrate could be made in this way for reconstitution with added water.

There are more agreeable-sounding uses for UF, now in common use: the purification of paint in paint baths, or its recovery from wash waters, and the concentration and purification of antibiotics and similar products from broths. In all these applications, UF provides an economic method of recovering the valuable material not only in a concentrated form, but in a purified form also. In many of these applications it also performs a vital service in aiding serious effluent disposal problems.

The flow sheets for such applications vary widely, but generally they provide for a very high recirculating flow through the plant, so that the concentrate flow may be up to 100 times that of the product flow. Applied pressures are usually below 10 bar.

Cleaning the membranes

Without the high liquid velocities obtained by recirculation, the process would come to a stop very quickly because of accumulation on the surface of the membrane. Even with the cleansing effect created by high turbulence, all these processes depend on regular backflushing and cleaning of the membranes to maintain normal performance levels.

It is usually sensible to establish with pilot tests just what cleaning methods and chemicals will have to be used, and what regime will have to be practised. The full-scale plant can then be built to incorporate the necessary equipment, such as backflushing lines and dosing points for cleaning chemicals. On some plants the whole cleaning cycle (which can be quite complex) is made automatic, like the regeneration cycle of an ion exchange plant. This makes sense if the cleaning process has to be carried out frequently, for example daily.

The small-diameter tubular membranes have the advantage that they can withstand pressure applied both from the inside and from the outside, because their support is integral with the membrane itself. This means that they can be backflushed by applying clean water under pressure on the outside of the tubes: the water passes through the substrate and the working skin, and loosens deposits from the inside of the tube. Large diameter tubular membranes can be cleaned with sponge-ball pigs which are a tight fit in the tube; water pressure forces them down the tube so that the surface of the membrane can be wiped clean. With suitably chosen membrane materials, quite aggressive chemical conditions can be used for cleaning, including acids, alkalis, disinfectants and enzymes, together with detergents.

The object of pilot trials is to establish operating and cleaning regimes which will allow the membranes to work for three years or longer before being replaced, a lifetime similar to that of RO membranes.

UF in water treatment

In the process applications mentioned above, UF can perform duties for which no other process is suitable. In water treatment this is not the case; we can use UF instead of RO or membrane filtration because it seems economically more favourable. On the other hand, water treatment applications are, on the whole, less demanding than process applications; UF will operate at higher fluxes, lower pressures and with less frequent and rigorous cleaning regimes.

The duties which UF might perform in water treatment include the removal of suspended matter and colloids, of large organics and of bacteria. RO will do all these, and remove most of the dissolved salts as well but at a higher cost. Membrane filtration will remove all these, except the large organics. The main field in which this kind of removal is needed is in the production of ultra-pure water, above all for the manufacture of microchips and microelectronic components, and also for other industries such as pharmaceutical manufacture.

The classical flow sheet (if we can so describe a process which is only a decade old) for ultra-pure water production is RO–DI–membrane filter. (A dirty water may need pretreatment before RO, and exceptionally clean waters may not even need RO at all.) The final membrane filter serves to remove particles and bacteria which have slipped through the previous processes or been generated in them; sometimes there is a central membrane filter at the water purification plant outlet but point-of-use filters are commoner. They are always installed for demanding applications.

In this classical flow sheet, UF might replace the RO ahead of the deionizers but UF will not of course remove any dissolved salts, so that the DI plant would then have to cope with the full raw water salinity. Alternatively, UF could be used as a final polisher. Can UF perform either of these duties better than the classical alternative? In some cases, it appears, it can.

UF in ultra-pure water production

Where the water has a low TDS, the economic advantage of having most of the salts removed by RO is small. UF will perform the same kind of duty with respect to removing the large size impurities and will do so at a much lower pressure. If the water is relatively clean, then the flux can be up to five times greater than that on a spirally wound module, so a much smaller membrane area has to be installed for the same flow-rate. Pressures tend to be lower than in process applications, at perhaps 5 bars. In suitable applications these economies will be greater than the additional capital and operating costs which are incurred by the deionization plant having to remove the entire TDS of the raw water.

The most promising use for UF in ultra-pure water production, however, seems to be as a post-filter after the deionization plant where it is installed for two potential benefits:

- As UF has a much smaller pore size than any membrane filter, its use at this point will result in a better final water quality. Experience shows that UF actually seems to raise the resistivity of the water, which is not easy to explain.
- The other benefit is that UF plant has a modest and predictable operating cost, especially in this position as post-filter, where it is fed with very clean water. Membrane filters, on the other hand, are throwaway items which have to be fitted with new cartridges when clogged. This can lead to high operating costs if, for some reason, the water is high in filterable materials.

UF does not replace point-of-use filters, which are still necessary as final safeguards, but would act as a central polisher after the water purification plant. In some cases it is claimed that the installation of such a UF plant has paid for itself, in reduced filter cartridge costs, within a year.

We have just come across two areas of ignorance: how a UF plant can increase the resistivity of a water, and why there are some plants which produce a highly purified water which is still high in filterable materials. We have to be clear in our minds about this kind of thing: when working in the realms of ppb and sub-ppb concentrations of impurities, we are very often in the dark. As we are operating at the limits of our methods of analysis, this is hardly surprising. In this kind of work there is, as yet, no substitute for experience and trial-and-error.

Pretreatment for membrane processes

We can now review the three membrane processes described in these chapters. One common factor emerges strongly: membrane processes are above all governed by the occurrence and prevention of fouling of the membrane surfaces. The secret of satisfactory operation in all three processes is therefore a satisfactory raw water quality and a suitable regime of membrane cleaning. In most cases this means that some form of pretreatment is required to protect the membrane plant.

RO plant in particular should never be installed without at least a 5 micron filter to remove coarser suspended particles. Tests using membrane filter discs will establish the fouling index or SDI of the water, and will indicate whether further pretreatment is needed. Dissolved or suspended iron in the feed water is one of the most damaging foulants; most of it is removed by ion exchange softening, so this is a very common form of pretreatment for smaller-scale plants. It is reliable and simple, and on the small scale the cost is modest.

Flocculation and coagulation is a more efficient form of pretreatment but the process is impracticable on small-scale flow and especially on intermittent operation. It is normally restricted to large potable water plants. If imperfectly operated it can produce a water full of unreacted aluminium salts and fine $Al(OH)_3$ floc, which can be more damaging than the untreated water would have been.

Flocculation/coagulation plant is always followed by sand filters; sometimes the sand filter on its own is sufficient. Other pretreatment options include ion exchange scavengers and activated carbon filters. Even UF itself can be used to protect an RO plant.

RO treating natural water usually needs to be dosed to avoid scale formation. Acid and/or anti-scalants can be used; a reasonable body of experience is beginning to build up in this respect but we still rely heavily on trial-and-error.

With so many options available, it is difficult to decide on the best combination of security and economy. In each case we have to consider the interaction between the nature of the raw water and the desired quality of the treated water, the type of membrane module to be installed, and the scale of the plant and its patterns of utilization. It is in resolving such complex problems that skilled design and operation show their true worth.

Part 8
ULTRA-PURE WATER

22 General

Ask the average person about ultra-pure water and they will say brightly, 'Ah yes! distilled water'. Forty years ago they would have been right. In those days we didn't use much high-class water; what little we did need was produced expensively by distillation, and might have contained about 1ppm of total solids.

A whole new technology of water purification has developed since then, with processes like ion exchange deionization and reverse osmosis. The general advance in water purification technology now allows us to produce water down to about 1ppb of total impurities. (Parts per million (ppm) is more precisely defined as milligrams/litre, and parts per billion (ppb) as micrograms/litre, but for brevity we shall use ppm and ppb.) It is fair to say that modern industry depends on our ability to produce ultra-pure water, without which present-day electricity generation or the production of microelectronics would both be impossible.

Ppb is easily said, often by people who have no feeling for what they are really about. It means one part of impurity per 1,000,000,000 parts of water; in terms of time this ratio is the same as one second every 32 years; not a bad watch! The last paragraph says the total impurities in it are 'about 1ppb' because we must be cautious in our statements here. These purities are on or beyond the limit of accurate detection, where nothing can be said with real certainty. *No other substance, natural or man-made, exists in anything like this purity. Anything to do with ultra-pure water is in a class of its own.*

Who needs it?

The impurities in water include inorganic salts, organic materials, suspended and colloidal matter, and living organisms. To some extent, of course, these categories overlap; for example, bacteria are also suspended matter. Different users lay stress on the level at which each of these categories becomes important to them. There are three main users of ultra-pure water.

- The electricity generating industry is the biggest. A typical high pressure generating set takes about 1500 tonnes an hour of the stuff, converts it to steam, passes it through the turbine, condenses it and uses it again. We will go into some detail about high-pressure boiler feed water and methods of polishing the recycling condensate to keep the feed within specification. However, its purity specification is concerned almost entirely with non-volatile inorganic impurities. Organics are undesirable but less critical, and microbiological growth can hardly be important in a boiler.
- The pharmaceutical industry is also a high-purity water user. Water for injections must be pyrogen free, which effectively means it must contain no large organic molecules, to ensure that pyrogens (a group of large molecules resulting from the decay of living organisms) occur at concentrations of less than about 10ppb. (The actual criterion for pyrogenicity is simply whether or not an injection of the water will raise the body temperature of a rabbit, but the actual concentration of pyrogens which will bring about a serious reaction is about 10ppb.) Total sterility is of course essential, but relatively large amounts of inorganic salts can be tolerated; for example the water may contain up to 1ppm (1000ppb) of chloride.
- The electronic component industry, particularly in the manufacture of microchips, needs the purest water of all. In effect it demands a purity at the limits of detection for all classes of contamination: inorganic, organic, suspended and colloidal, and microbiological. Whether one class of these impurities is particularly harmful, or why, is unknown; all that the industry knows is that by insisting on total purity, the percentage yield of satisfactory chips is maximized whereas the slightest contamination in the wash water causes the yield to drop.

At this point, it is interesting to look at the kind of impurities with which we are now particularly concerned. Dissolved materials are fairly easily removed by ion exchange, and suspended matter by filtration; the great difficulty in ultra-pure water lies in that no-man's land between suspended and dissolved matter. Another is the exclusion of microorganisms and suppression of their growth. Figure 22.1 shows the sizes of these materials in relation to the other impurities found in water.

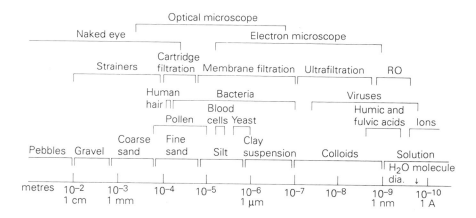

Figure 22.1. Contaminant size spectrum

How to make ultra-pure water

Making ultra-pure water is relatively easy. It can be assured by the use of a train of successive purification processes which include a reverse osmosis stage, followed by two or more stages of deionization. This flow sheet is increasingly being augmented by final ultra-filtration, or even reverse osmosis, to meet the high degree of security needed by some end users.

Some waters are easier to treat than others and do not need the full rigour of such a flow sheet. For example, a clean water such as we might expect from a deep well may not need the RO stage to remove particulates. The deionization stages may be a pair of single bed units, followed by a mixed bed, or two mixed beds in series. It all depends on the quality of the raw water and on the degree of security which the end user demands.

Problems can arise if the raw water cannot readily be treated by reverse osmosis, for example if it is badly contaminated with iron and/or organic materials. Then it must be subjected to pretreatment, not so much for the sake of product purity but rather to protect the RO membranes. RO treatment more or less guarantees that its product will contain only traces of suspended or colloidal matter, or of organic molecules above a molecular weight of about 200. In theory RO also yields a water free from microorganisms.

The mixed beds which follow reverse osmosis take out the remaining dissolved inorganic salts. This is where we have possible problems. The removal of Na^+ ions is typical of all the other ions to be removed, and will therefore serve as an example.

Ion exchange is a reversible process and the removal of Na^+ on the cation exchanger goes in both directions:

$$Na^+ + Re.H^+ \underset{\longrightarrow}{\leftarrow} H^+ + Re.Na^+$$

The resin has a strong preference for Na^+ ions, but this is not absolute. The reaction which takes out Na^+ ions goes from left to right, and it will be the predominant one, but we can never completely suppress the reverse reaction from right to left, which puts Na^+ back into the water.

The rate at which the unwanted reaction takes place depends on the reaction constant (which is a property of the cation resin and not within our control) and on the availability of the two reagents which take part in it. If there were no Na^+ on the resin, and/or no H^+ in the water, that is if the concentration of H^+ and/or of $Re.Na^+$ were nil, then the reaction from right to left could not take place and we should theoretically get a product water containing zero Na^+. Chemists perform dreary mathematical operations to demonstrate this, but a commonsense view is quite good enough to show us what the issues are. If we want to make ultra-pure water, free from Na^+, then we must either treat a water without any H^+ ions in it or use a resin without Na^+ on it.

H^+ ions occur in cold water at the very low concentration of 10^{-7}, so when we are looking for a product water of ordinary purity we can say the availability of H^+ ions is so low that we can call it zero. In ultra-pure water this is no longer true, and the concentration of H^+ is fixed beyond our control. The only fruitful course of action left to us is to eliminate $Re.Na^+$ — that is, avoid having any Na^+ ions on the resin.

Again, in commonsensical terms, this means that the final polishing mixed bed must be as fully regenerated as possible, so that its capacity is all in the form $Re.H^+$ and none in the form $Re.Na^+$. It is uneconomical to achieve this by using massive quantities of regenerant to try and get total regeneration of the resin. The best method is to have a final mixed bed which has never received any significant quantities of Na^+ ions; then the problem of taking

Na$^+$ off the resin will not arise. That is why the process train must contain two or even three ion exchange stages in series, so that the last one never receives any significant load at all, but is nevertheless regenerated regularly to make quite sure that no Na$^+$ accumulates on it.

However, a mixed bed can also acquire Na$^+$ by imperfect regeneration techniques. While the cation component is being regenerated with acid, the anion component receives caustic soda, a copious source of Na+, and a very damaging one if it is allowed to come into contact with the cation exchange resin. If the separation of the resins is imperfect, or if the centre distributor is at the wrong level, then trouble is bound to follow.

This need for perfect regeneration has stimulated the development of three-resin systems, in which an inert resin layer is introduced to classify between the cation and anion components, and thus ensure perfect separation. These systems are not always trouble-free, and have to be installed and used with care and expertise. When they work well, however, they provide an excellent solution.

Keeping pure water pure

Producing pure water is the simpler part of the problem. Nature abhors ultra-pure water, just as it abhors a vacuum. From the moment it is made, the water is at risk from re-contamination of every kind.

Obviously the entire pure water circuit must be completely isolated from the outside world, and contained in equipment whose materials of construction do not allow anything to be leached out. The design and construction of ultra-pure water circuits is a specialized art which will be touched on later.

Fighting off the bugs

The chief difficulty in ultra-pure circuits is to prevent microorganisms breeding in them. No impurity can be prevented absolutely from getting into the circuit, and that includes bugs. Bugs are different from non-living material because they can breed *in situ,* without further reinforcements from outside. In theory, one bug is enough. One may well wonder what nourishment they can find in an ultra-pure water circuit; unfortunately they will always find enough.

We speak of sterilization, but what we usually do is to disinfect, that is, we kill off the vast majority of organisms in the system but cannot guarantee to kill them all. If we do this regularly and efficiently, we shall have a near-sterile system. There are several methods we can use: steam is the most lethal to bugs, but can only be used in all-stainless steel systems; UV irradiation is an effective disinfectant but it only works in the place where it is applied, and it leaves no residual biocide in the water (biocides normally have to be rinsed out because they are unacceptable for whatever use the water is put to); very fine filters or membrane processes theoretically hold back all bacteria but the odd one will always get through. Sterilization, therefore, is not straightforward and, because it is so important, we normally have to use a combination of methods to ensure purity in the circuit.

It is essential to construct the circuit in such a way that bugs find it hard to settle anywhere and also such that it can be efficiently disinfected.

Monitoring purity

The conventional method of defining ultra-pure water is by its electrical resistivity, especially in the electronics industry. The specification has risen progressively from a resistivity of 1 megohm (20 years ago) to 10, then 15, and most recently 18 megohms. If we consider that the theoretical resistivity of pure water at 25°C is 18.2 megohms (as determined by Kohlrausch, hence the name Kohlrausch water for the theoretically pure material), this is tantamount to saying that the theoretical resistivity is the required value; certainly there is no room for a control band.

There is no immediately obvious merit in resistivity as the prime parameter. In practice it is used because conductivity measurement is relatively simple, reliable and cheap (the important word is relatively because at these extreme levels even conductivity meters become delicate and complex). Conductivity will not measure microbiological impurities or colloids and particulates. However, if we succeed in achieving Kohlrausch water at 18.2 megohms, it is a good bet that we have removed everything besides dissolved salts.

The drive for higher resistivities can be taken to absurd lengths; this sort of conversation has been known to take place: 'We want you to guarantee a resistivity of 20 megohms.' 'But the theoretical conductivity is only 18.2 megohms!' 'If you want to tender, you have to guarantee 20 megohms!' The plant supplier cheats: he guarantees 20 megohms and adds (in fine print) 'at 20°C'. Nobody gets much benefit from this.

Some very sensitive (and expensive) instruments are being developed to supplement conductivity measurement: particle counters, total organic carbon meters, bacteria detectors; even the LAL test for pyrogens (see p.235) has been used. All these are still imperfect, which is not surprising. For the present, resistivity measurement rules, but we must recognize the difficulties which it, too, encounters at extreme purity.

The obvious difficulty is that the resistivity of pure water is extremely high, so that the signal available for monitoring will be small. The less obvious difficulty is the change of resistivity with temperature. At conventional levels of purity, conductivity is used to measure the dissolved salts in the water. Most of the current is then carried by salt ions and the conductivity is therefore roughly proportional to the ionic content of the water. The variation with temperature is then a function of the water's resistance to ion transport, which decreases by about 2 per cent per deg C, which makes temperature compensation a simple matter.

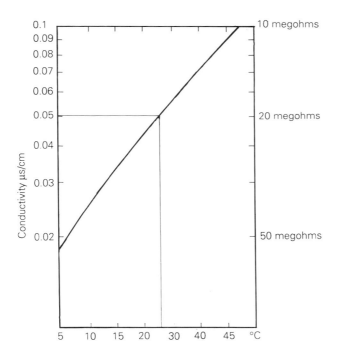

Figure 22.2. Variation of conductivity of pure water with temperature

In Kohlrausch water, the current is carried only by H^+ and OH^- ions. These are present in tiny concentrations, but in practice we are looking for salt ion concentrations which are still smaller. Changes in temperature not only change the resistance of water to ion transport but also change the dissociation constant of the water itself, so that the number of H^+ and OH^- ions available for carrying current will increase with rising temperature. Figure 22.2 shows the variation of resistivity of pure water against temperature. The compensation needed is not only much greater than 2 per cent per deg C but the change is not linear. Accurate measurement requires either a controlled sample temperature of 25°C or elaborate temperature compensation built into the instrument.

Summary

In ultra-pure water we are largely working in the unknown. When product quality specifications are at the limit of detection, we have no control band. The effective solutions to problems in this field owe less to scientific rules than to a body of good practice which has only evolved over very recent years. To achieve satisfactory operation in practice depends on three equally vital components: the plant must be designed on sound principles, and must have a generous safety margin at every point; the ultra-pure water circuit must be made of suitable materials and design, with high-class workmanship; and the production of water of consistently high quality needs first-class housekeeping.

23   Condensate polishing

Why should we want to clean up condensate? If we feed a boiler with good quality water and then operate it in such a way as to minimize carry-over, the steam should be even purer than the boiler water and therefore the condensate ought to be fit to put straight back into the feed.

This argument holds for the great majority of boilers, but there are two classes of exceptions: a medium- or low-pressure system in which the condensate becomes seriously contaminated; and, more importantly, boilers requiring ultra-pure feed, that is high-pressure and/or nuclear power-station boilers.

Contaminated condensate

Most uses for steam involve closed heat exchangers in which the steam condenses without becoming contaminated. With a reasonably well-designed system and reasonably good housekeeping we can avoid excessive grease or oil contamination, and adventitious pick-up of impurities in the condensate return system. Condensate does indeed pick up corrosion products from the piping and heat exchangers, but if this is going to prove a serious problem in the re-use of condensate for feeding medium-pressure boilers, then even more serious damage is probably occurring somewhere along the line and such a condition should not be allowed to continue.

On the other hand, there are a few industries in which the steam inevitably becomes contaminated, or at least where there is a high risk of this. In sugar

213

refining, for instance, condensate derives from multiple-effect distillation plant in which the syrups are concentrated. Any leakage in this equipment, or priming, or carry-over of the syrup will contaminate the condensate with sugar. (In fact, there is no convenient method of removing sugar from condensate and the proper remedy is to monitor the condensate quality and be prepared to dump, if necessary.)

Pulp and paper mills produce a condensate rich in suspended matter (much of it corrosion products), which can be filtered out. Petrochemical condensate may suffer from oil contamination, against which sand filters are installed, though some forms of oil are exceedingly difficult to remove.

Each of these problems is an individual case, and little can be said about them which is useful as a general rule. The condensate must be considered as an alternative source of feed water; if it can be brought up to a suitable quality more cheaply than raw water, well and good. If it cannot, there may be some alternative use for the water. In a large steel mill, for example, a supercritical boiler is fed with freshly deionized make-up only, and its contaminated condensate is then used as make-up to the mill's medium-pressure boilers.

High-pressure boilers

On the question of the modern high-pressure boiler, there are two factors which point towards condensate purification.

In the first place, the steam/water separation becomes less efficient as pressure increases. At 10 bars pressure, the density ratio of water to steam is 150:1. At 180 bars (a normal pressure for large modern drum boilers) the ratio is only 3.7:1, and at the critical point and beyond there is no steam/water separation at all. (At 374.1°C and 221 bars, steam and water both have a specific gravity of 0.318.) Therefore, as we approach the critical point, the boiler ceases to concentrate and retain impurities in the water; an ever-increasing proportion will pass on with the steam. Boiler blowdown no longer works as an efficient kidney for the circuit.

This holds true for all impurities, but with silica the problem is even worse. Silica is soluble in high-pressure steam and little will be retained in the boiler even at 180 bars. Silica is one of the most troublesome contaminants in turbine generators; it comes out of solution in the turbine and forms a glassy, adherent layer on the blades, which progressively throttles the steam flow and ultimately leads to out-of-balance running, which causes dramatic failures.

The other problem is that in order to maintain its efficiency, the high-pressure boiler needs ultra-pure feed water whose total contaminants should not exceed about 5µg/l.

The trend towards 'once-through' boilers, that is without an effective drum, accentuates the quality problem, as the contamination which enters the boiler must either pass out with the steam or deposit on the tube surfaces, because no blowdown of concentrated contaminants is possible.

The sources of contamination in the circuit are corrosion and dissolution of the surfaces of the boiler circuit itself, and cooling water ingress in the condenser.

Contamination from the circuit

When the boiler is first commissioned, massive quantities of rubbish are released into the steam/water circuit. However great the care in construction, the water will bring down oxides, swarf, oil and grease, and silica (probably from sand-blasting residues). More picturesque impurities, such as long-forgotten sandwiches, spanners, etc. may also materialize.

For an initial time, therefore, the boiler is cleaned and flushed, without coming on full load. Some users install temporary plant to help in cleaning up the condensate during this period, which is either rented from a contractor or put aside for the next boiler to be commissioned.

To a lesser extent, there will be a flush of impurities into the circuit every time the boiler is re-started. This is particularly serious if the boiler and the steam circuit have been allowed to cool down, when the surfaces face a completely different corrosion regime. However, even if the boiler is maintained near its working temperature, the change of temperature and of flow patterns will release deposits into the circuit. The need to purify recirculating condensate therefore becomes more acute in boilers which are not on continuous base load.

A new power station is obviously going to be more efficient than its ageing predecessors in the same system. It is therefore normal to run the newest stations on base load and so obtain the maximum benefit from their greater efficiency. In turn, these stations age and are superseded; at that point they may be relegated for use only to supplement the daytime peak demand, and be banked down for the night when consumption is low. It then becomes likely that, on a daily shifting regime, condensate quality cannot be maintained and condensate polishing plant has to be added where none may have been necessary before.

Contamination arising in the boiler circuit

Once the boiler is in full continuous working order, the rate at which circuit contamination from the surfaces enters the condensate will have stabilized. To a large extent the rate depends on the materials of construction and the boiler water regime.

Figure 23.1 is a schematic flow sheet of a high-pressure boiler and turbine generating set. Much of the corrosion takes place in the feed heaters; as the condensate from these heaters returns to the main circuit, we have to take into account corrosion on their steam side as well as the water side. Conventionally, the LP heater tubes are of cuprous alloy, and the HP heaters are steel. If the pH of the boiler water is low, then there will be corrosion on the steel HP heater tubes. If the pH is raised by ammonia (or by amines which become degraded to ammonia) then the steel will be protected but the cuprous alloy tubes may suffer ammonia attack. All the available cures, higher class alloys in the heaters, for example, are expensive.

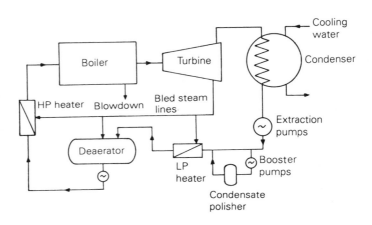

Figure 23.1. Schematic diagram of typical HP boiler/turbine system

Boiler design and operation therefore have to compromise and be prepared to suffer some corrosion. In any case, some oxide formation in the boiler itself is inevitable; at the operating temperatures, the tubes form an oxide film, some of which will be scoured off into the steam and water.

The physical and chemical form of these products in the condensate is indeterminate. (Generally, they have acquired the name 'crud' said to stand for 'corrosion residue and undetermined detritus', though less polite derivations are also current.) There will be some dissolved $Fe(OH)_2$ (bearing in mind the total absence of oxygen), some colloidal magnetite Fe_3O_4, and some particulate rust Fe_2O_3, together with traces of other metallic oxides. Their total level in feed water should not exceed 1 μg/l.

Contamination from the condenser

Efficient use of fuel, according to the laws of thermodynamics, requires us to exploit the energy released over the widest possible temperature range. On the one hand this has driven boiler pressures and temperatures up to their present high levels, with steam temperatures sometimes as high as 540°C. On the other hand, the low pressure end of the turbine must exhaust at the lowest temperature attainable with the available means of cooling. To achieve this, power stations use large and efficient condensers at high vacuum. Sea water provides a lower temperature coolant than a recirculating cooling system with an evaporative cooling tower. This is one reason why power stations are built at the coast rather than inland. In some areas of the world where water is in short supply, there may be a need to use air-cooled condensers; a consequence of this is that returned condensate temperatures are high, limiting the range of condensate treatment processes that can be applied.

In a sea-water cooled condenser, we have about 30 000mg/l NaCl in the cooling water at low pressure on one side of the tubes, and condensate of less than 2µg/l of Na under high vacuum on the other side; the ratio of Na concentrations of the two streams is 6 000 000:1. Even a minute weep of cooling water into the condensate will therefore cause gross contamination, whereas a ruptured tube will result in immediate emergency conditions.

Needless to say, these condensers are designed with great care. In the UK, for example, condensers are fitted with double tube plates and the gap between them is flushed with condensate to guard against the effect of a weeping tube joint. Increasing use of titanium as condenser material reduces the danger of contamination but imposes a significant increase on the capital cost of the installation. Nevertheless, the designer must take suitable precautions against greater or lesser failures.

Summarizing the problem

Condensate polishing in power stations therefore has to meet three quite different requirements:

- it must deal with the flush of impurities (dissolved, colloidal, and suspended) experienced at primary start-up and, to a lesser extent, again at subsequent re-starts;
- it must also deal with the wide range of normal corrosion products produced at a far lower level during routine operation; and
- it must be able to cope with cooling-water contamination, consisting essentially of dissolved salts and, in the case of severe failure, high concentrations of these.

Of the three duties, the first two are more or less predictable. The level of contamination due to cooling-water ingress against which the system is designed to guard is largely an arbitrary value, decided on at the specification stage.

The polishing plant has to treat a water which is already extremely pure and to yield a product which might typically contain less than 2μg/l of Na and less than 5μg/l of SiO_2.

A 500MW set will consume about 1500 tonnes of feed an hour, which is equivalent to a flow of water sufficient to supply a town of 150 000 inhabitants. From this massive flow of water the polishing plant has to remove teaspoonfuls of contaminant. That leads to some curious design problems.

Practical solutions

These problems are relatively new because in the field of large power-station design, progress and experience come more slowly than in other branches of technology. Power stations, like elephants, are massive beasts with long gestation periods.

The post-war improvements in steam technology produced a new generation of power stations which were built from about 1960 onwards. The size of single generating sets increased dramatically, and both temperatures and pressures rose far higher than ever before. The boiler drum became progressively smaller until, in a small proportion of boilers, it logically disappeared altogether. In these 'once-through' boilers all the feed is converted to steam, without the facility of blowdown to remove feedwater impurities in a concentrated form; whatever goes into the boiler must come out through the turbine. A further logical extension of this trend was to go to supercritical temperature and pressure, when no blowdown is possible anyway.

The new steam conditions were only possible because ion exchange deionization (then a new process) had made cheap and pure make-up water available. As the new stations came into use, it was discovered that high quality make-up alone would not suffice. To maintain the desired feed-water quality it was found necessary to clean up the condensate, too. At first this was introduced rather through the back door, by recycling a part of the condensate flow through the make-up plant. Next, the final mixed bed of the make-up plant was specially increased in size to accommodate this additional task. Finally condensate polishing became a process in its own right.

This immediately raised strategic questions. How much of the condensate needs polishing? Is it sufficient to take a part of the condensate and polish it, or is it really necessary to polish 100 per cent of the huge flow of feed water consumed by a modern generating set? If, indeed, it is necessary to polish 100 per cent of the flow, does it have to be polished 100 per cent of the time or can one economize by installing a single 100 per cent polishing unit which can be shared between more than one generating set? Moreover, should the polishing unit consist of a single vessel which must be bypassed during regeneration, or is it more prudent to go to the cost of splitting the flow between two 50 per cent units?

To some extent, the answers to these questions are dependent on the design of the generating set and its operating conditions, but to a large extent they are matters of policy rather than technical and economic optimization.

Tactical problems

The strategic questions having (we hope) been settled, we come to the tactics of dealing with the polishing duty. This duty consists of two functions: the plant must filter out solid materials ranging from colloidal size upwards, and it must remove all dissociated materials, from heavy metal salts to sodium chloride. Here again, the options available are numerous:

- whether or not to filter crud out separately?
- at what point, and which stream of condensate, should one polish? For example, one of the main sources of ferric oxides is the HP heater condensate: should special plant be used to deal specifically with that?
- should treatment be 'on line' or 'dump and return'?
- what type of filter/ion exchange system to apply?
- what level of standby plant is needed?

Filtration — a renewed concern

Both resin and equipment manufacturers in the ion exchange business have always worried about getting particulate matter on to the resin beds, and when the first condensate polishing plants using ion exchange resins were being considered back in the 1950s and early 60s, so did the boiler makers and utilities. Examination of the published literature of the period shows that the accepted wisdom was that crud transport into the boiler system was at least as dangerous as ionic transport, and perhaps more so.

It would therefore have been surprising if the early condensate polishing plants had not included some attempts at filtration. If, however, one looks now at the profusion of condensate polishing plants of different types installed worldwide which make no attempt at pre-filtration prior to ion exchange, it might be concluded that the condensate polishing pioneers were playing it very safe, were ignorant or were dealing with a problem that in fact did not exist. With hindsight one can see that all these are at least partially true and as, for

219

reasons which will become clearer later, condensate filtration is again being considered seriously it is worth reviewing some at least of the history of filters in condensate systems.

Restricting our interest to the modern power-producing unitized boiler–turbine–condenser system, it should be apparent that most, if not all, of the particulate matter in the system is generated internally since the only liquid addition to the system is the make-up water, which is not only highly purified but amounts to only a few per cent of the system recycling flow. So any crud in the system has to be generated by corrosion or precipitation mechanisms, or perhaps somebody left it there in the first place. (Unlikely though it sounds, this last mentioned is not unknown. The case of the bag of sand in the boiler drum is well known, and rumours of a bag of cement in another boiler drum are well founded. This in addition to the usual pile of welding rods, old boots, etc.)

Although there has been a move away from copper-bearing materials in such circuits, they were extremely common in the 1950s when the standard condenser-tube material was admiralty brass. Cupronickel components in the feed heating system were common, and indeed are to this day. With air pumps on the condensers, and full-flow heating deaerators further up the heating system, there was plenty of classical practical chemistry and operating practice to define a sound operating regime. Keep the pH value above, say, 8.8 with ammonia, dose some hydrazine for residual oxygen scavenging and also make sure the pH does not rise too far above, say, 9.2 to protect your copper alloys, and you should be in a situation where you will have minimum corrosion of both ferrous and non-ferrous components. For lower pressures the hydrazine might be replaced with sulphite, and the boiler drum might be dosed directly with caustic soda as an insurance against condenser leaks upsetting drum chemistry; but by and large, under steady base load conditions, the corrosion arising in such a circuit should not be large, and so it has proved.

First commissioning and restarts were, of course, a different matter. One would expect there to be some offload corrosion, particularly of the large area of carbon steel, and there were definite worries about the amount of crud which might be transferred to the drum during such times because there was, and is, only a limited amount of water that can be dumped from the system before the make-up plant becomes severely over-strained.

What was unavailable, at least to the majority of the people involved at the time, was a more detailed understanding of the types of iron oxides involved in the corrosion mechanism, and this ignorance had some consequences which were at least unfortunate.

The initial view taken was that ion exchange plants in condensate polishing should be protected by some form of pre-filter, and it was argued that the filters might actually be more valuable than the ion exchange equipment.

Sand filters

Filters in the 1950s were shallow-bed sand filters operating at a specific flow rate of, maybe, 10 metres per hour. Condensate flows are large; a quick rule of thumb for modern high-pressure units is 3 tonnes condensate per hour per megawatt, so such an installation was going to be huge. By comparison, incidentally, the ion exchange plant might be rated at, say, 100 metres per hour, and thus occupy 1/10th of the area.

Moreover, sand filters contained silica, and the desirability of having lots of silica in high-pressure boiler feed circuits was questionable. Nevertheless, tests revealed that at normal condensate temperatures achieved in UK practice the silica pick-up was negligible, and the first condensate polishing mixed beds in the UK were preceded by a large bank of vertical-pressure sand filters, which also had facilities for precoating with a cellulosic powder material.

Pre-coat filters

Now if a precoat could be used successfully on a sand filter, it appeared obvious that there were better ways of employing it. By the 1960s, there had become available quite a range of precoat filters, both in tubular candle form and in various leaf forms, either vertical or horizontal, with all the manufacturers making the usual claims of superiority for their particular equipment. Most of these precoat filter designs had in fact been developed for other industries such as food, pharmaceuticals, brewing and mining where the precoat material usually used was diatomaceous earth, which is another good source of silica; but the cellulosic precoat material looked like a good bet for power-station work and laboratory tests, using synthetically prepared particulate iron, revealed no problems either in filtration during the operating cycle or in cleaning filters.

Thus quite large numbers of precoat filters of various types were ordered in many parts of the world for condensate polishing duty, most of them using a combination of stainless steel filtration elements with the cellulose precoat material.

These systems were almost universal failures. It had not been appreciated that the range of iron oxides which can be formed in power stations is extremely large and that not all of them will filter out readily. Moreover the amount and type of oxide will vary with the operating regime and the standard analysis methods for total iron do not tell you which type of oxide is present. The black Fe_3O_4, magnetite oxide, filters fairly easily; the yellow/brown and red hydrated oxides may barely be filtered at all. Much worse than this, however, was the discovery that some of these oxides would penetrate the precoat materials and form a layer with the cellulose next to the filter septum which proved almost

221

impossible to remove with the usual *in situ* clean-up technique. In extreme cases, the elements had to be removed manually for cleaning every 2 or 3 cycles. Operators soon started bypassing the filters and found that the ion exchange units themselves seemed to suffer no great harm; indeed they worked as filters at least as well as anything else and usually a good deal better. Moreover the cellulose contributed variable amounts of sodium and organic matter to the treated water.

In the intervening years we have learnt that deep beds of ion exchange resin at flow rates up to 300 metres/hour are indeed excellent filters of power-station crud and a variety of clean-up techniques has been employed to remove this crud from the resin, with marked success. As a general rule, these techniques are the opposite of the conventional make-up plant practice, in that suspended material is removed downwards usually by a short sharp rinse, following an air scrub. This procedure is beneficially applied both before and after chemical regeneration. Ultrasonic cleaning has also been used successfully.

Resin particles

Sulphate is now recognized as an important and highly undesirable contaminant in boiler feed water. Its reduction to levels of less than 2ppb may be required, and some authorities even want to reduce it to levels of perhaps 0.2–0.1ppb. Cation exchange resin is sulphonated polystyrene and, if a resin bead splinters, it is quite feasible that these splinters will go through the bottom collecting systems and the resin trap systems; in fact such systems are designed to allow this to happen to prevent their becoming clogged up. If you want to work to these very low sulphate residuals, it might seem wise to prevent spoiling your wonderful ionic purity by letting slip a piece of broken resin which might become sulphate ion in the boiler, and we can predict that much finer filters after the ion exchange plant in condensate polishing will become common in the more sensitive installations. Indeed, one operator of large pressurized water reactors (PWRs) has already installed 5 micron filters in this position and, in Japan, huge UF systems have been installed in spite of their high capital and operating costs.

Two other systems for removal of particulate matter, the magnetic filter and powdered resin filtration, are worth comment.

Magnetic filters

Some forms of oxide are highly magnetic, and in particular magnetite can be removed quite readily by a sufficiently strong magnetic field. Moreover, a magnetic filter has one very clear advantage over all other types of filters: its ability to withstand very high temperatures, thus allowing its positioning at the boiler inlet. On the other hand it will not be particularly effective at removing some of the hydrated oxides, and the commercial magnetic filters available have so far proved extremely expensive in capital cost. Their impact has thus been rather muted, in spite of their apparent advantages.

Powdered resin

Powdered-resin filters represented a major innovation in the field of both filtration and ion exchange. Standard-size ion exchange resins of cation and anion types occasionally show a tendency to clump if mixed together, particularly when brand new. If the individual resins are ground down to a fine powder and mixed, they expand dramatically to form a material which can be used as a precoat. This material has been shown to have excellent filtration properties, noticeably more effective than the cellulosic materials, and particularly so when dealing with hydrated oxides. It will also, of course, have ion exchange properties but, because of the very small weight of ion exchange material present in the circuit, the total capacity is limited and also, since there is no way of recovering the material for regeneration after use, its operating cost will be high if the ionic load is much more than a trace of contaminant.

This filtration system has a particular niche in boiling-water reactor (BWR) systems where the exposure of the fuel cans directly to the circulating heat exchange media renders them more susceptible to crud fouling. Moreover, the volume of spent material to be dealt with on recoating is quite small and, since BWR systems are by their nature radioactive, the waste-handling charge may be noticeably lower than in other systems. The neutral pH of the BWR also suits the ionic removal role of powdered resins.

Ion exchange in condensate polishing

The preferred design of condensate polishing plant is a function of the chemical operating regime of that particular system. To allow more detailed examination of why a particular design is chosen, it is useful to consider the components of the preferred operating regime in rather more detail. These may be stated concisely (not in any particular order of importance) as:

- Ionic purity of final treated water with particular respect to undesirable ionic species. These are usually taken to be sodium and chloride. Sulphate has been added in recent years, and the inclusion of other criteria can be forecast.
- Preferred chemical conditions in condensate. This is reflected primarily

in the condensate pH, which is easily elevated into the range pH8.8–9.6 with ammonia, or is neutral in the case of BWRs or boilers operating under VGB (Technical Association of Large Power Plant Operators — Germany) guidelines. However, addition of other constituents has to be taken into account, for example hydrazine, morpholine or other amines.

- Likely concentration of undesirable species in the raw condensate, due to condenser inleakage on the water side. The biggest single contributor here is likely to be water-side condenser leaks, particularly in sea-water cooled installations.

- Accidental inleakage of gases from the section of the feed heating train under vacuum. Although the condensers will be fitted with an air pump system of some description, returned drains into the condenser may not be brought back into a position where they are degassed very thoroughly; in any case condensers are not generally designed as vacuum deaerators although they obviously function as such to some extent. Thus CO_2 from atmospheric inleakage can be a serious and often under-rated problem.

- The contribution of contaminants from the make-up plant. As the make-up water is generally extremely highly purified, and the percentage make-up is low, this is not normally considered a serious problem. However, non-reactive silica and organic matter entering the feed system from this source can be expected to come under increased scrutiny in the future.

- The mechanical/hydraulic design of the entire feed heating system and, in particular, the heater drains return system. For example, in a typical fossil-fuelled power station, the high-pressure heater drains will be routed to the deaerator under normal operating conditions, but may be diverted to the condenser on start-up. The LP heater drains are usually returned to the condenser but other systems exist. In the PWR, however, the heater drains may be forward pumped so that the polishing plant installed at the classical point downstream of the condenser extraction pump will only see a limited percentage of the total system flow.

- Finally, all the above conditions will need to be examined, both with respect to the normal flow and the abnormal conditions which may or may not obtain during start-up and commissioning.

Examining some of the criteria outlined above in greater detail, one can note that the quest for improved ionic quality has been extremely rapid over the last two decades. In the mid 1960s, specific ion analysis for levels in the sub-ppm range was in its infancy, and the only online instrument of sufficient accuracy and reliability for general use was the conductivity meter. At a reference temperature of 25°C, a conductivity of 0.1µS per centimetre is equivalent to 18ppb of caustic soda or sodium chloride, and about 10ppb of hydrochloric acid (both expressed as calcium carbonate).

These were lower levels than could be measured by instrumentation for the specific ions at the time, and indeed to this day the conductivity meter remains a most valuable and rugged purity indicator. It is, however, only an analogue of purity and, since the variation in the specific conductivity with temperature differs for the different components, its value becomes more limited as quality requirements become more rigorous.

The first instrument to reach online maturity for condensate quality checking was the sodium ion electrode system which, initially, allowed reasonable confidence at levels in the 5ppb Na range. Since sodium carryover was known to be highly undesirable in boiler turbine systems, sodium limits in treated condensate appeared quickly, initially at the 5ppb Na level, and fairly soon reduced to levels of 1–2ppb Na for the more sensitive systems.

In the late 1970s and 1980s, the advent of the ion chromatograph has allowed really low concentrations of many different contaminants to be measured, of which the most important have so far been chloride and sulphate. The detection limit for these species is now in some cases better than 0.1ppb, and such ionic limits are starting to be demanded in some systems. This chapter is not meant to contain a detailed description of why any particular system demands a particular quality but, as an illustration, the case of the well-known Westinghouse PWR may be cited.

The earliest Westinghouse steam-generator designs were not considered to be particularly vulnerable, and the operating chemical regime was, typically, a co-ordinated phosphate treatment without condensate polishing. These steam generators were fitted with a blowdown treatment plant, usually of about 1 per cent maximum continuous rating (MCR) capacity, whose principal reason for existence was to cope with any primary-to-secondary circuit leaks, along with any crud which might have gathered at the bottom tube plate face. This regime was justified by the rather low steam conditions, temperatures and heat flux rates. These, however, increased in subsequent designs, and experience showed that these generators were perhaps more sensitive than had at first been anticipated. Criteria were eventually fixed by the manufacturer which effectively set a limit of 100ppb of sodium in the steam-generator water, together with an approximately stoichiometrically equal figure for chloride. With only 1 per cent blowdown available, there existed an effective demand for a 1ppb Na maximum concentration in the feed water to the steam generator.

In the UK, the Central Electricity Generating Board (CEGB) and the South of Scotland Electricity Board (SSEB) originally specified a 5ppb Na limit for the advanced gas-cooled reactors, later requirements being 2ppb for sodium, chloride and sulphate, with tighter limits being at least welcomed if obtainable, and possibly demanded in times to come.

225

As far as feed-water conditioning chemicals are concerned, the principal changes over the last 20 years or so have been the general acceptance of ammonia/hydrazine as the preferred choice, with a gradual increase of pH, and thus ammonia concentration, which parallels the increasing use of ferrous components in the systems at the expense of copper-bearing alloys. Notable exceptions are the neutral pH BWRs where the circulating water is exposed directly to strong radiation, precluding the use of such conditioning chemicals, and the neutral pH high-oxygen system advocated by the VGB which promotes magnetite growth for corrosion prevention by deliberate injection of oxygen into the feed system, coupled with a very low circulating water conductivity.

With the above in mind, we can now look at some of the systems which have been used for condensate polishing over the past two decades or so, and what might be applicable over the next few.

Mixed beds

It was not in the least surprising that the mixed bed was chosen for condensate polishing at an early stage, and has remained in the forefront of this field through to the present day. Mixed beds were, after all, designed to produce very high water quality and were particularly suitable for dealing with trace amounts of contaminants in raw water. Early experiments showed they could be operated at high flow rates, say 120 BV/hour without apparent kinetic problems, so that it was perfectly feasible to arrange for the typically large condensate flow to be treated in a bank of operators, moving the mixed resin from each operator to a central regeneration stage when necessary. Although specific flow rates were quite high, the ionic loads were generally low and the run times thus quite long, typically a week or thereabouts under normal conditions.

The difficulty that had to be faced was that mixed beds are awkward to regenerate. The resins have to be separated, and especially with *in situ* regeneration systems it is never possible to achieve the ideal 100 per cent separation of the two components. With the added complication of a resin movement system, the difficulties in achieving this multiplied since the volume of resin in any one operator was unlikely to remain constant and the proportion of cation to anion might also vary, so that the regeneration system was being asked to cope with two new variables.

The consequence of all this is that some of the cation resin sees caustic soda during regeneration, and some of the anion resin is exposed to sulphuric or hydrochloric acid. Thus the final mix contains more sodium on the cation resin than one would prefer and more sulphate or chloride on the anion resin than is theoretically ideal.

The view taken initially was that this would not matter too much because the equilibrium values needed to obtain the purity of water required were not crucial when operating in the hydrogen form. For example, to obtain an outlet quality of less than 5ppb Na and 5ppb Cl in the hydrogen form mixed bed, the cation resin typically must have a minimum of 24 per cent of sites in the hydrogen form, and the anion resin 6 per cent of the sites in the hydroxide form. If this resin phase composition is not achieved, then the output values will not be reached, irrespective of condensate quality at the inlet. Indeed, the plant will actually add contaminants to the water until equilibrium is achieved. Incidentally, even if the required output values are theoretically possible, they may still not be obtained if the ion exchange zone is insufficient for the removal being attempted.

In the absence of any extreme operating conditions, the rather crude early condensate polishing mixed beds were capable of achieving this level of performance, but in an unpublished paper presented to the Filtration Society in 1971, it was demonstrated that a mixed-bed polishing plant of this type with 3-year-old resin could not cope with a continuous condenser leak of approximately 2ppm (as $CaCO_3$), while brand new resin could.

This sort of observation was not terribly encouraging when looking at operation in the hydrogen form, but if one considered the implications of operating in the ammonium form, when ion exchange equilibria are much less favourable, it was positively terrifying. The advantages of operating a mixed bed in ammonia form were persuasive. The major load being removed during normal running is ammonia, which has been deliberately added to the system. The cation resin will remove it, so it has to be replaced. The economic advantage is quite clear. However, to achieve the 5ppb sodium and chloride levels mentioned before, for a typical resin operating at a pH of 9.6, one can afford only 0.4 per cent sites in the Na form on the cation resin, and 3.8 per cent sites in the chloride form on the anion resin. It was obvious that sodium values of less than 5ppb were going to demand special measures, and various methods were sought to achieve this.

Powdered resins

This is an appropriate point at which to consider the advantages of the powdered resin system for ionic removal. Since the resins are, by definition, brand new and highly regenerated, they should have no particular problem in meeting such requirements, even when operated at high pH. To some extent experience with operating plants has confirmed this. They do, however, have their own difficulties in meeting sub-ppb level requirements. First of all, the regeneration level has to be exceptionally high, and the standard resins from which the powdered resins are manufactured will not necessarily be at this level of purity as received at the resin grinding plant. This was always known to be a problem with the anion resins but could also be true for the cation resins if one was looking for figures down to the 0.1ppb Na figure, where the purity of ammonia used for ammonization would also become crucial. This would increase the manufacturing cost to a point where the resin became hopelessly uneconomic, and it must be remembered that, since it is dumped after use, the operating economics of the powdered resin system have always been marginal at the best of times. There is also a mechanical problem in that, if 100 per cent of the septum area is not precoated, then quite clearly a small proportion of the water passing through will not have been treated at all, and this proportion does not have to be very large when one is aiming at these high qualities.

For these reasons, the powdered resin system has not achieved the pre-eminent place in ultra-high purity condensate polishing which at one time looked likely. As stated previously, its specific advantages in radioactive applications ensure its continued place in the condensate polishing plant repertory.

The requirement of 1ppb of sodium, or better, demanded solution, as did the achievement of similarly low qualities for anionic species, and at this point one can note two different approaches (which are not mutually exclusive): the application of chemical solutions to the problem; and mechanical/hydraulic solutions.

Chemical solutions

Since the initial concern was that of sodium on the cation resin, and since the prime source of this sodium was from the caustic soda applied to the anion resin, one could look at chemical methods for removing the sodium from any cation resin entrapped with the anion resin before returning the system back to service. For example, in one system, the anion resin and its entrapped cation resin is purged with a low concentration of ammonia which may then be recycled through the separated cation resin in the cation resin regeneration system. A new equilibrium for sodium will be achieved throughout the total cation resin in the system, and regenerating the separated cation resin only with acid will then ensure a significantly lower total of sodium contamination than otherwise. This system gained quite wide acceptance, and in good operating conditions can achieve sodium values about the 1ppb level at elevated pH. It does not, however, address the problem of anion resin contamination by the acid regenerant.

In another process the regenerated anion resin and its entrapped cation resin are treated with lime slurry in which the high calcium ion concentration will reduce the sodium ion concentration on the cation resin to very low levels. The calcium ion is strongly held, and this should not give problems in subsequent operation, provided of course that the volume of the entrapped cation resin is not too high; but the process does somewhat suffer from a degree of emotional antagonism.

Cation-mixed bed

It may be apparent by this stage that inadequacies in the mixed-bed regeneration technique could lead directly to the addition of impurities (particularly sodium and sulphate ion) to the treated water. Obviously the less frequently the mixed bed is regenerated, then the lower the total weight of the contaminants added to the system over a given period, although the actual concentration might be the same in any one regeneration.

In a neutral pH system such as BWR, the ionic load in the raw condensate must be close to zero in the absence of condenser leaks, so a mixed-bed plant will require infrequent regeneration; this possibly explains why ionic contamination problems in BWRs have received much less attention.

If the system chemistry requires a high pH value, then the ammonia (or other amine) will be the predominant load, and the run length between regenerations will be relatively short.

In these circumstances, one could consider installing a hydrogen form cation unit ahead of the polishing mixed bed, the sole duty of this plant being to remove ammonia, so that the mixed bed normally has very little load with which to deal and can have an extended operating period between regenerations. Incidentally, it happens that the cation unit will probably remove the majority of crud present in the condensate circuit, which will do no harm either to the mixed-bed operation.

Although this system is more expensive in capital cost, it has clear attractions and it should be noted that plants of this type in the UK have shown themselves capable of long-term production of excellent treated water qualities in the advanced gas cooled reactor (AGR) stations, in spite of the fact that the mixed beds, which were designed back in the late 1960s, did not have particularly sophisticated regeneration systems or sequences.

Mechanical solutions

Mechanical methods emphasized initially the avoidance of the sodium problem by arranging the interface, or the resin separation and sluicing point level, such that some anion resin was entrained deliberately with cation resin, but not vice-versa. As concern with anionic contamination grew, this crude approach was replaced by more sophisticated systems. In a further system, a high strength caustic soda was used to float the anion resin away from the entrapped cation resin.

Various companies appreciated that the resin at the interface after separation was a trouble spot, and simply removed this 'trouble resin' from the system during regeneration, storing it in some separate container so that it took no part whatsoever in the regeneration procedures. It was logical to follow this by the substitution of an inert material when such became available. Initially these inert resins were allowed back into the operators, but in some cases their affinity for oil or other coatings occasionally present in condensate caused subsequent embarrassment, and typically they may now be used only in the regenerators.

The ultimate attempt at the mechanical solution may be the abandonment of the mixed-bed principle altogether and its replacement by separate beds of cation–anion–cation, avoiding the cross-contamination problem in its entirety. Such systems also have additional advantages in the achievement of improved kinetics over mixed beds, but at the price of enhanced capital cost.

In the late 1980s, the position can be summarized by stating that the best of the mixed-bed systems using closely graded resin, possibly with an inert intermediate, and careful control of resin volumes throughout the operation cycle can achieve levels of better than 2ppb of sodium, chloride and sulphate at elevated pH values and sub-ppb values when operated in the H^+-OH^- form.

Resins — the choices available

Bearing in mind that in almost all condensate polishing plants the resins are moved from the operators to the regenerators and then back again, the question of the physical strength of the resins has always been considered important. The various manufacturers of ion exchange resins have always understandably promoted the best features of their particular product so that, for example, manufacturer A extolled the resistance to abrasion and osmotic shock of his macroporous resins, while manufacturer B pointed to the superior performance of his gel resins in the Chatillon test and manufacturer C claimed that all his beads were of identical size, etc. There were considerable differences in preference from country to country, and indeed from individual to individual, but the situation may be summarized reasonably even-handedly by saying that the best of the gel resins, both cation and anion, now available have at least adequate strength for condensate polishing duties, while their generally higher ionic total capacities and operating capacities are usually advantageous, and in some cases essential. There have been marked improvements in the last few years in the availability of such resins with a much improved uniformity coefficient, to the extent that we may now be approaching the ideal situation where resins can be specified at a precise bead size with remarkably little in the way of variation either above or below the best requirement for separation and/or maximum kinetic ability. And this without paying dramatic premiums in resin cost.

This is highly satisfactory, but it should also be recorded that considerable care still needs to be taken in ensuring that resins used for condensate polishing, and indeed other duties associated with sensitive power plants, are free of manufacturing residue. Cases of emergency shutdowns, due to high chloride levels in steam generators from breakdown of organochlorides, are not unknown.

Recent publications reveal yet another problem: cation resin is a sulphonated polystyrene. Short-chain molecular fragments from the cross-linked matrix can arise from careless manufacture, or from chemical degradation of the matrix. Such fragments are water soluble and get leached out in the course of the run. They are also anionic, and effective anion resin foulants. They tend to clog the anion resin pores and slow down the rate at which ions can travel in and out of the resin. The effect is especially marked in the anion resin's ability to remove sulphate ions, which are particularly large and slow to move. Both cation and anion resins are now being improved to minimize this effect.

Some mechanical considerations

The best regeneration method in the world, coupled with 100 per cent resin separation, will still be inadequate if one fails to remove all the resin from the operator unit at the end of the operating cycle. It is easy to obtain 95 per cent resin removal, and 98 per cent is not too difficult, but achievement of 99.9 per cent demands much more care in bottom system design. As a general rule any mechanical 'clutter' at the bottom of a unit will provide hiding places for resin, and this should be avoided. Of the systems that are known to give really excellent resin removal two may be mentioned.

First there is the flat wedge wire screen covering the entire bottom surface of the operator vessel. Resin movement is best via a single point at the centre of this screen going downwards, and removal of the final amounts of resin may be assisted by a circumferential sluicing ring, or by deliberate inducing of a tangential swirl.

Similar designs with flat nozzle plates are likely to prove difficult, but if the nozzle plate itself is dished, and a single outlet at the centre is employed, almost complete removal can be obtained in conjunction with a sluicing ring or similar system.

It has been practice in the United States, and indeed in some other countries, to specify that the bottom collecting system be designed to a differential pressure equivalent to the system operating pressure, and this may be extremely high.

The booster loop system in which a local recirculation pump of, say 105 per cent MCR capacity is used to replace the head loss in the condensate polishing plant (see Figure 23.2) has many advantages. The flow rate through the condensate polishing plant is kept constant irrespective of the operating regime of the power generator, which not only makes it easy for the operators but avoids changes of flow which have been shown to be likely to interfere with filtration efficiency. If a booster loop system is fitted, it can be argued that the bottom collecting system need only cope with the maximum head capable of being generated by the booster loop pump. This can represent a significant saving in capital cost on the condensate polishing plant design. It should be noted, however, that the safety care design of some nuclear systems precludes the open bypass line which really forms an intrinsic part of the booster loop concept.

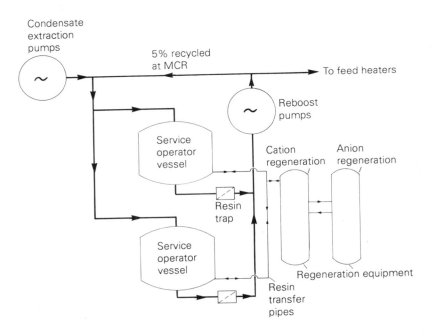

Figure 23.2. Schematic diagram of typical condensate polishing plant with loop recycle and external regeneration

24   Pyrogen-free (PF) water

We have pointed out that the use for which the water was wanted made great differences in our approach to its production and handling. There are respects in which water for pharmaceutical production is quite different from water for microchips or for high-pressure boilers.

Pharmaceutical specifications

We have to consider various pharmacopoeias whose requirements vary slightly. The most important of these are the British (BP), European (EP) and the US Pharmacopoeia (USP). Each of these specifies various grades of water, of which Aqua Purificata in the BP (and its equivalent in the others) is not particularly pure but reflects the quality which an old-fashioned still could produce. The specified limitations on the individual inorganic contents are all well within the scope of a simple deionization plant. The only exception is that some specifications limit narrowly the pH, such as the USP's pH limit of 5.0–7.0. In a high-quality deionized water pH is so difficult to measure that in practical terms this is meaningless. Moreover, mere contact with the atmosphere could cause enough harmless CO_2 pick-up to bring the water outside specification. The water is also required to be sterile, which is a problem to be discussed below.

    We get into ultra-purity when dealing with pyrogen-free (PF) water. We have to recognize two grades here: Aqua ad Injectabilia or water for injections (WFI), which is the only grade which pharmacopoeias allow for actual use in preparations destined for injections. In addition, pharmaceutical manufacturers use large flows of PF water for ampoule washing etc., but without having to obey the full rigour of the pharmacopoeia's regulations for WFI.

In the EP WFI is defined as water used in the preparation, dissolution or dilution of medicaments for parenteral administration, that is for injection directly into the bloodstream.

The specification for its preparation, quality and distribution includes the following:

- The water must be prepared by distillation, starting with potable water, purified water or distilled water. The distillation apparatus must be made of neutral glass, quartz or 'suitable metal' and be fitted with a device to prevent the entrainment of droplets.
- For distribution in bulk it must be filled into glass containers (or 'other materials complying with the country in which the product is used'), heat sterilized and sealed.
- The chemical tests for the water are the same as all those specified for purified water. These are not particularly stringent; for example, the residue after drying at 100–105°C must not be more than the equivalent of 30mg/l. In addition it specifies a test for pyrogens, in which the water is made isotonic with NaCl and injected into rabbits at 10ml of sample per kg of the rabbit's body weight. At least three rabbits must be used and their temperature taken regularly for three hours. The rules for disqualification of individual rabbits, interpretation of results and repeat tests required in negative or doubtful cases are complex and make the procedure very costly.

It is apparent that producing water for pharmaceutical use is simple except when it comes to WFI or PF water. This chapter will therefore deal primarily with pyrogens and their removal.

Pyrogens

As early as 1865, it was observed that injections into the bloodstream frequently led to elevated body temperature, but the cause was only identified in 1923 as 'a filterable heat-stable substance of bacterial origin'. This pioneering work established that freshly distilled water is pyrogen-free but that it would become re-contaminated on standing unless it was kept sterile.

Many substances can act as pyrogens but materials of bacterial origin, mostly gram-negative bacteria, are the only ones of practical importance. They are a group of complex lipopolysaccharides with molecular weights typically in the region of 20 000, whose exact chemical nature is not known. Their virulence varies and measuring their concentration as such gives no reliable indication of their effect. To show the orders of magnitude involved, we can say that a dose as low as 0.002μg per kg of body weight injected into the bloodstream can cause a detectable rise in body temperature. If we are to allow a reasonable safety margin, therefore, the concentration of pyrogens which can be tolerated in water for injections is of the order of 5 micrograms per litre.

The variability of the agent and the very low concentrations involved mean that the main difficulty in producing and maintaining PF water is in the monitoring.

The only definitive test admitted by both the BP and EP is injection into a group of carefully controlled rabbits of samples of water and then the monitoring of their temperature. This is expensive and, by the time it shows a result, a large quantity of product may have been prepared with the water or the water under test may well have become re-contaminated. As a method of direct plant control it is impracticable.

As pyrogens are organic materials, they must contribute to the total organic carbon (TOC) level. As one might have expected, good correlation has been found between TOC levels and pyrogenicity. However, TOC measurement is at present insufficiently sensitive; to become useful it will have to be able to detect levels at least an order of magnitude lower than at present.

The LAL test is sensitive to the common pyrogens and is widely used for practical control but it is not ultimately accepted in law. The test relies on an extract from the blood of the horseshoe crab, which clots in contact with minute concentrations of these pyrogens. The intrinsic difficulties of the test are the standardization of the extract and the detection and definition of clotting in the sample but techniques have improved rapidly since the test was first introduced in the early seventies and it has even achieved some measure of automation. At present it is our nearest approach to a rapid and convenient pyrogen test. It is highly sensitive in detecting the only common pyrogens, which are those derived from gram-negative bacteria. It is therefore widely used as an indication that the water is pyrogen free.

Storing and transporting pyrogen-free water is difficult so PF water is preferably produced at the time and point of use. This implies at least some small local production units which can not support the cost of intensive supervision and maintenance. Because for practical purposes we have no convenient and reliable method of testing the water on a continuous basis and because such small units have to be able to operate reliably without high-level attention, the production methods have to be totally reliable.

It is largely the absence of a reliable 'pyrogen meter' which causes the EP and BP to lay down that WFI water must be prepared by distillation, a process which is so reliable that instant and constant monitoring are unnecessary. These regulations of course apply equally to large-scale pharmaceutical manufacture, where the level of supervision which is economically possible means that this simplicity and reliability is not so essential. As a matter of interest, WFI is the only product in the pharmacopoeias for which not only the maximum impurities but also the route of manufacture is specified. There are, however, alternative processes available.

Processes for removing pyrogens

Membrane processes such as reverse osmosis (RO) and ultrafiltration (UF) will reliably remove large organic molecules. RO is already in wide use for producing PF water whereas UF has only recently been introduced for this type of duty and is not yet fully proved. They do not, however, give the same degree of security as distillation.

The reliability of distillation is for two reasons: any maloperation which might allow pyrogens to pass into the products would immediately cause high conductivity due to dissolved salts coming over with the organics; more importantly, WFI stills are designed to yield the product at 80°C. The water is then kept at this temperature until it is used, which prevents bacteria breeding in it and producing pyrogenicity.

On the other hand, distillation is expensive, especially at today's fuel costs, even though modern multiple-effect or vapour recompression stills have greatly improved the efficiency of the process.

RO presents a much cheaper alternative. Its membranes reject all materials of molecular weight above (about) 200, which includes all pyrogen molecules and it does this at a fraction of the cost of distillation. However, the water is produced at ambient temperature and the clean side of an RO module is a perfect site for colonization by bacteria. Re-contamination can therefore take place before the water even leaves the plant. It is uneconomical to heat the water to 80°C only to cool it down again. With distillation, by contrast, maintaining the water temperature at 80°C is more or less free. The use of RO for PF water production thus calls for the highest quality of supervision and housekeeping, to ensure that those bacteria which will inevitably get round to the clean side of the plant are destroyed before they have the opportunity to multiply.

The USP and some others, unlike the EP and BP, already allow RO to produce WFI, but subject to stringent controls. It is likely that the EP will eventually follow suit and there is already considerable pressure from pharmaceutical manufacturers for it to do so.

In pharmaceutical production, the volume of PF water which is required for ancillary use, like ampoule washing, is normally far greater than the volume used in the injectable preparation itself, which must be of WFI grade. This ancillary stream can legally be produced by RO. It has therefore become common practice to install RO for the larger flow of PF water, working side by side with a still to meet the needs of the injectable preparation itself. Such a dual system leads to some interesting flow-sheet possibilities.

The LAL test is widely used to monitor these streams. It is not theoretically acceptable to the inspectorate for the control of WFI, but is useful in monitoring whether quality is being maintained, especially in PF water production.

Producing pyrogen-free water

The pyrogen-removing process, whether distillation, RO or (in the future, perhaps) UF, will always be the last process in a purification system which normally consists of a number of consecutive processes.

Distillation

The efficiency of a modern distillation plant in terms of water usage and heat consumption depends on a high-quality feed water, so that the feed can be almost wholly evaporated without causing corrosion or scaling. Pharmaceutical product stills are normally constructed in stainless steel and fed with a good quality demineralized water or condensate. Softened water can be used, but only if the total dissolved solids of the raw water are low.

Condensate may only be used for feeding WFI stills if it is free from noxious boiler additives such as filming amines. It must be free also of significant carry-over of incondensibles from the boiler. The dangers of poor quality feed to WFI stills are illustrated by one well-documented case when a hospital pharmacy's triple-effect stainless steel still was fed with condensate from the main steam system to the hospital. The boilers appear to have primed so heavily that even the product from the still occasionally failed the Aqua Purificata specification for chlorides. Worse than this, the concentration of chlorides which built up in the water remaining in the still was sufficient to corrode the stainless steel beyond repair within months of its installation.

Reverse osmosis

RO only rejects about 90–95 per cent of the inorganic salts in the water and by itself it will not therefore produce inorganic impurity levels which are below the Aqua Purificata specification. RO must therefore be preceded by a deionization plant which will remove most inorganics. This need not produce a particularly high-quality water; on the contrary, some polyamide hollow fine fibre membranes hydrolyse if fed with very pure water. A feed at a conductivity of $10\mu S$ is usually satisfactory.

With a feed so low in inorganic salts a relatively low reject flow from the RO can be tolerated. Even then, RO is rather wasteful of water and is liable to reject 30 per cent of the input water. Because the feed is already rather pure, the RO reject stream represents a significant flow of high-quality water which can be put to good use, such, for example, as boiler feed. In some flow sheets, it is used as feed to the WFI still but some authorities consider it bad practice to feed stills with a water in which organic residues have already been concentrated.

RO will only operate on very clean water. Suspended or colloidal matter will foul the membranes and so will large organic molecules of moorland waters. If the raw water is rich in any of these, successful operation of the RO plant depends entirely on the installation of a suitable pretreatment plant and its careful and consistent operation.

Maintaining quality

All pharmaceutical production involves preventing bacterial growth and this is particularly critical for PF water. This is because once bacterial growth has been allowed, the water will be contaminated with pyrogens, even if the bacteria are subsequently destroyed.

In this connection, it is best to use the word disinfecting rather than sterilizing to remind us that we can kill most of the bugs most of the time but can never be sure of killing all the bugs all of the time. The most positive method of disinfection is with steam, and therefore pharmaceutical water systems are generally built to be capable of withstanding high temperatures, in stainless steel, and equipped with connections to permit regular steam sterilization. RO plant cannot be subjected to steam sterilization, hence the doubts regarding its safety and reliability.

Means of disinfection in widespread use also include:

- inorganic biocides, primarily chlorine and hypochlorite, but peroxide and ozone are being considered more commonly in recent years;
- organic biocides, of which formaldehyde is the commonest.

These two add something to the system which cannot be tolerated in the product water but which would maintain a residual disinfecting action. When such biocides are used, the whole system must be flushed before normal product can be resumed. Formaldehyde is particularly difficult to flush out and is therefore used only when other biocides are inadvisable. Peroxide and ozone are somewhat different in that they offer residual disinfection for only a short time but then decay harmlessly to water.

Then there are:

- ultra-violet (UV) light irradiation; and
- membrane filtration, to 0.2 microns

which are capable of being efficient disinfectants if carefully installed and maintained, and do not add anything to the water. On the other hand, they leave no residual disinfecting capability. They are installed as in-line devices but both are fallible. UV effectiveness falls sharply if the water contains any suspended or colloidal material which shields the bacteria. Filters collect debris and thus provide a nutrient-rich site on which other bacteria can grow. There is also a good deal of debate whether it is possible for bacteria to 'grow-through' the pores of a filter. The ideal system combines UV and filters in the final circuit.

Practical systems use a combination of two or more of these methods. For example, continuous disinfection by UV and/or membranes is generally backed up by regular sanitization with steam or biocides. Recent work recommends the use of continuous ozone generation within the final circuit, with UV lamps at the point of use, which destroy any residual. The actual choice rests with the type of system and, in particular, its materials of construction. It can be a difficult one.

Pyrogen-free water systems

It is most important that the system should be designed first to make it hard for bacteria to grow in it, and second to make disinfection simple and efficient. Many aspects of the design and operation of PF water systems are shared with ultra-pure systems for microelectronics, and the specialized guidelines which apply to both will be discussed in Chapter 25.

The main difference between PF and other ultra-pure systems is that, in PF water, relatively high levels of inorganic contamination are allowable. Piping and storage systems can be built in stainless steel, rather than in plastic, because the minimal amount of inorganic matter which tends to be leached off the former does not affect the product quality. Stainless construction permits the use of regular steam-sterilization.

Operation and maintenance of PF systems

If all the guidelines for plant design and construction recommended are observed, the resulting plant ought to operate with a minimum of trouble. Very few plants in the field approach this ideal, partly because of costs but above all because PF water systems are usually the result of years of changes, additions and adaptations. The greater the departure from these guidelines, the greater will be the need for careful housekeeping and maintenance.

When trouble does occur, some elementary errors and negligence usually

239

come to light: storage vessels which ought to be vented through bacterial filters have been exposed direct to the atmosphere; UV lamps have been allowed to fail or to become covered with deposits; systems which normally rely on their 80°C operating temperature for protection have been allowed to cool down during shutdown and not re-sterilized. With hindsight, all these are simple errors but without painstaking supervision they are inevitable.

Conversely, some plants whose design appears to break every rule in the book operate satisfactorily for years, due presumably to the care with which they are operated. Producing PF water is not a science but a craft, where skill and experience can overcome material deficiencies.

25 Water for the electronics industry

Taking stock of the technology described in this book indicates that equipment and processes are now available to us to produce very pure water indeed. This section concentrates upon the use of ultra-pure water by the electronics industry.

The contaminants which must be removed can be broadly classified as soluble in water and insoluble (although some insoluble particles are so small that this classification can be questioned), with additional sub-division into organic and inorganic. To add further complication, bacterial growth can, and does, occur in very pure water, and this contamination must also be removed if the product is to be defined as ultra-pure water.

The flow sheet of a modern plant can usually be considered to consist of three sections. The centre section will contain ion exchange equipment, probably preceded by reverse osmosis (RO). Before the water enters this section, it may be pretreated to remove material which would either prevent the ion exchanger and membranes giving a long, trouble-free life or, worse still, pass through to the final polishing section. This final polishing part of the plant consists essentially of the storage and distribution equipment, and will contain some method of bacterial control and ion exchange polishing, as well as fine filtration.

Pretreatment

The extent of pretreatment has all too often in the past been dictated by cost considerations rather than technical requirements. Ideally, the pretreatment plant should present a clean water, that is, free from all suspended matter, containing the minimum organic material, to the next stage of treatment.

Feed water may be derived from a number of sources. It may be a river, a lake, a borehole or a mixture of all three. In Chapter 1 we outlined some of the impurities likely to be encountered with these various sources. The simplest plant will contain only filters, but these, even multi-media types, are not cure-alls and should do what the name implies, that is, filter. If large soluble organic molecules are present, they will not be filtered out nor, indeed, will insoluble material consisting of particles less than 2 microns. A chemical addition, such as aluminium sulphate or an iron salt, is very effective in bringing about flocculation in these circumstances. Another chemical treatment required at this stage is chlorination (and dechlorination) to reduce the bacterial count as well as to oxidize some of the larger molecules present. Activated carbon is effective in the dechlorination role but has a limited life as an adsorbent. An organic trap, that is an anion exchange resin regenerated with alkaline brine, is effective in removing those organic materials which would otherwise foul a downstream strongly basic anion exchange resin working in a demineralization plant. Organic scavengers, however, do not remove 100 per cent of the organic matter present.

Demineralization

If the feed water is a surface supply, it will often contain more organic matter even after pretreatment than water from a deep borehole. A membrane system will help reduce the organic contaminants. If the feed water is derived from a borehole, then a membrane system in the form of an RO unit will also significantly reduce the inorganic ionic solids. An increase of the temperature of the water before the RO modules will reduce the capital cost of the membrane installation. Depending on the type of membrane process selected, some chemical conditioning of the water will also be necessary.

Ion exchange, preferably counterflow regenerated cation and anion exchange units, may also form part of this section, so that a pure water as free from contaminants as possible is fed to the polishing section.

Polishing section

This is the area where the emphasis should be on the maintenance of the water purity. Correct choice of materials, no stagnant areas (to minimize bacterial activity), ultraviolet light (to reduce the bacterial count), prevention of air-borne contamination by nitrogen blanketing, fine filtration by membrane plant and correctly designed and installed control systems must all be considered as priorities in this area by any competent design engineer.

Flow sheet

A typical flow sheet is shown in Figure 25.1. The character of the feed water should be the main consideration of the designer when considering equipment in the pretreatment and demineralization sections. The polishing section equipment is independent of the feed water quality. Flow sheet experimentation will be a feature of the coming years, since the availability of variations on a theme will be readily apparent. For instance, there are designers who would like to reduce the ion exchange equipment inventory by the use of RO in series. Whatever flow sheet is employed, the high cost of making ultra-pure water makes preservation of purity mandatory. Much of what follows in this chapter will therefore deal with the means of keeping pure water pure.

Ultra-pure water systems

The demineralization section which produces the pure water generally includes an RO step, whose output is continuous, and which should not normally be operated on a stop–start regime. The first sections of the plant are therefore usually rated to meet the demand for pure water, which is variable, with continuous or near-continuous production.

This means that the purified product must be accumulated and stored in a storage tank from which it has to be conveyed with its purity unimpaired to the actual points of use. There are usually a number of points of use, and they may be some distance away from the central water purification plant, requiring a complex piping system. This piping system is usually in the form of a recirculation loop. Impurities will build up in the loop, and the power loss due to pumping will warm the recirculating water. The materials of construction of tanks and pipes must not leach or otherwise contaminate the water, and the main fight is against bacterial growth in this extensive system, promoted by the increased temperature. Some form of polishing in the loop is essential.

Materials and construction

The electronics industry's demands for inorganic purity are too stringent to allow the use of stainless steel, which releases metal ions and particles into the water; the use of plastic materials is mandatory for both pipes and tanks.

243

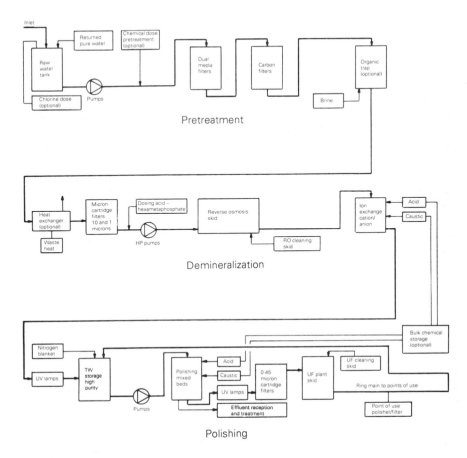

Figure 25.1.   Typical flow diagram for electronics industry

Plasticized materials must be avoided as the plasticizer is prone to leach. Unplasticized PVC, polyethylene and polypropylene, ABS or PVDF, have all been considered, and the choice between them is discussed below.

Pipe jointing can present a problem. For conventional duties PVC and ABS are easily joined by solvent welding, which is simple and can produce a good and clean internal joint. However, in high-purity work this operation is suspect because solvents may be held in the joint and escape slowly, possibly together with polymer residues. Heat welding is the preferred method for ultra-pure systems.

It goes without saying that the construction and assembly of the system must be carried out in clean conditions, leaving no grease, solvents, swarf or filings behind.

Steam sterilization of the whole polishing circuit is impossible; even polypropylene will not withstand the temperature and pressure involved. Sanitization of the system therefore has to use a chemical biocide, such as chlorine. This is not as positive as steam because it cannot be guaranteed to penetrate every crevice or reach live bacteria protected by a film of dead bacteria cemented by slimes. As a result, the choice of materials, design and workmanship of the system, and good housekeeping in operation, all become that much more important.

Plastics

The ideal plastic materials of construction should combine an intrinsic resistance to being colonized by bacteria with a very smooth surface on which it is difficult for them to lodge. ABS is prone to leach and is unacceptable for really high-quality work. Hydrophobic materials such as polyolefins are more easily colonized than hydrophilic and are not normally used.

This leaves unplasticized PVC (UPVC) which is hydrophilic and has the added advantage that it can be made with a specially smooth internal finish. (However, there are recent reports that this smooth surface breaks down in time and then gives a particularly good foothold to biological growth.) PVDF, a fluorinated polymer and a relatively novel material, is doubtless the best now available but it is very expensive. High-grade UPVC is still the most commonly used material and it may be some time yet before buyers will be convinced that the superior properties of PVDF are worth the extra cost.

Recirculating pipe systems

Bacteria can only grow on solid surfaces, and they cannot lodge on a surface while a brisk flow of water passes over it, so no part of an ultra-pure water system should ever be stagnant. The normal plant has a loop drawing from the pure water storage tank, round all the various points of use and back into the storage tank again. Flow is continuously maintained in this loop, with a recommended velocity of 2m/sec, minimum.

This counsel of perfection is impossible to meet under all conditions, and skilful design is needed to achieve a good compromise. When water is taken out of the loop at points of use, there will be a lesser flow of water coming back. Loops are therefore often designed with progressively smaller pipe diameters in the return section so that velocities are maintained when pure water is drawn off at the points of use. There must obviously be some kind of compromise which allows the whole flow to recirculate at times when no water is being drawn off.

In branches which tee off to the points of use, there is no flow when the take-off valve is shut. The valve should, therefore, go as near the loop as possible and so reduce the stagnant surface to a minimum. At most, the distance between the loop and the valve should not exceed four times the pipe diameter. In due course, all systems are subject to changes, additions and modifications. It is then important to leave no dead legs connected to the system, a precept which is often ignored.

Design of storage tanks

Storage tanks must, of course, also be perfectly smooth inside, with a conical or dished bottom so that they can be drained completely. Air vents must have a biological filter to avoid drawing in air-borne bacteria when the tank level falls. (Failure to check the condition of this filter is a common cause of trouble.) A system using a blanket of nitrogen is a more secure method of preventing air-borne contamination.

The tank should be sized to provide the balance between continuous operation of the RO unit and the intermittent demand for ultra-pure water. Too large a tank, however, increases the possibility of bacterial growth, and recirculation of purified water should always be considered a useful option on the flow sheet.

The contents of the tank should be constantly on the move, for example by bringing the inlet from the recycling loop into the tank at a tangent so that it promotes a swirling motion. However, the main site of bacteriological contamination is the dome of the tank above the water level, which is inevitably wet and has no flow preventing growth. Some authorities recommend that the inlet jet should point at the dome to keep it scoured. Perhaps a better way of avoiding growth in the air space is to install a small UV lamp at the top of the tank, although problems of attack on the tank surface, as well as lamp replacement, must be considered.

Maintaining purity in the loop

A recirculating loop needs a polishing plant to remove impurities which would otherwise accumulate in it. This usually includes an ion exchange mixed bed loaded with special high-grade resins which will not leach residues into the product, to maintain high resistivity in the water. As ion exchange beds are a favourite site for bacterial growth, they are normally placed after some means of catching any bacteria which might be coming round.

Ideally, this consists of a UV lamp to kill the bacteria, followed by a 0.45 or 0.2 micron filter, or even a UF module to collect the corpses; in practice either the filter or the UV lamp is often omitted for economy. An effective UV lamp alone theoretically prevents breeding in the circuit; the filter alone theoretically filters out all bacteria and might therefore be expected to keep them out of the circuit, but bacteria do get through filters in spite of the manufacturers' claims. If those bacteria which are retained on the filter are still alive they will breed on the filter surface or create spores which might penetrate through the filter. An additional UV lamp may be placed after the mixed bed to destroy anything breeding in the resin.

Sanitizing

Filters and UV lamps kill bacteria but confer no lasting protection; any organisms which get past them remain free to settle somewhere and breed. Biocides which would leave a residual in the water, such as chlorine, are of course unacceptable in ultra-pure water. It only takes one organism to get into the loop for it to settle and multiply there — with non-sexual reproduction, it doesn't even take two. All the precautions listed above therefore serve only to delay and minimize the inevitable outcome. The bugs will get through, and they will grow somewhere.

The system has to be taken out of service and sanitized at intervals. As this must be done with chemical biocides, it inevitably means lengthy flushing to purge the system of the biocide residues. The frequency with which systems are sanitized is therefore kept to a minimum.

Biocides include oxidants like chlorine (in some form) and peroxides, or formaldehyde. The oxidants are aggressive and may attack the plastic of which the loop is made (leading, for example, to breakdown of the smooth UPVC surface). The common oxidants damage ion exchange resins, hence the use of non-regenerable cartridges at the point of use to avoid having to sanitize them. Where an ion exchanger has to be sanitized, peracetic acid may be used but is expensive. Formaldehyde does not have these disadvantages, but is disagreeable to handle, and thought to be a health hazard. Its main disadvantage is that it is particularly difficult to rinse out.

All these chemical biocides are of little use if they do not get to the affected place. If the loop contains crevices or dead legs it is likely that bacteria will remain alive in them. Another possibility is that biological growth may have developed to such a point that the surface is covered by a layer of living and dead bacteria and their secretions. Such a layer will shield the organisms underneath from the biocide. A system can become so badly fouled that it becomes impossible to clean out.

Point of use polishing

For the best quality water, the last point before the water is actually used is equipped with a 0.2 micron filter, and perhaps a small cartridge mixed bed.

The future

The production and maintenance of water at 18 megohms is still a novel technology, relying on art as well as science. The main difficulty is the detection and analysis of fantastically low levels of impurity in the water. That involves costly equipment and manpower and there are only a few centres capable of carrying out reliable analyses at the required levels. Most ultra-pure water users do not have enough work to utilize such instruments and analysis fully, and this is likely to remain a characteristic of the ultra-pure water industry.

The feeling now is that modern pretreatment, RO and ion exchange processes are capable of removing all dissolved impurities to a satisfactory level, and that analytical techniques are capable of monitoring this. The removal of organic matter and fine particles, including living organisms, on the other hand still represents an area of weakness. The filters and membranes now available have been shown to be less than perfectly reliable, as are the methods of monitoring very small particles and total organic carbon. There is considerable room for improvement here.

The materials of construction available for ultra-pure water are either suspect or extremely expensive, so there is room for improvement here as well. On the other hand it does not look particularly likely that much better methods of sanitizing ultra-pure water systems will emerge.

There is a school of thought which says that plastic systems inevitably become bacteriologically contaminated in time, however good their design and housekeeping. Seven years is said to be the period beyond which there is no hope of keeping a system clean. Perhaps this view is based on the materials of construction available years ago. It may now be possible to extend the life of a pure-water system to a more economical period.

The demands made by the electronics industry are becoming more stringent. The gap between conductive paths on their devices is constantly being made smaller, with 0.1 micron gaps already in prospect. The dissolved inorganic impurities in 18 megohm water have been removed to near theoretical levels but we still have to improve water quality and its monitoring with respect to particulate matter, and we have to learn to cope better with the problem of biological growth. These are the changes for the future.

People sometimes ask how bacteria can get enough nourishment to grow in ultra-pure water. The answer is that they don't need a lot. A bacterium, at about 1 micron diameter, weighs about 10^{-6}mg. Then 100 bacteria per litre weigh the equivalent of 10^{-4}mg/l, or 0.1ppb. Most of that is water. The carbon required for this level of biological contamination is thus far below our powers of detection, that of the other constituents even lower.

Part 9
DEAERATION

26 The background to design

Dissolved O_2 or CO_2 in boiler feed causes corrosion. These gases must either be removed by chemical treatment in the boiler or by deaerating before the boiler (or both). Actually, most boiler feeds contain very little CO_2, so that O_2 removal is the main object of deaeration.

Deaerators and degassers

In the specialist jargon of boiler feed-water treatment, a deaerator removes O_2 from boiler feed and a degasser removes CO_2 in the make-up train. These are the definitions which we use; in other fields of technology the same words have quite different meanings.

The gas laws and Henry's Law apply equally to both processes. Table 26.1 shows the Henry's Law constant for O_2. When compared with CO_2, we find that O_2 is about one-thirtieth as soluble as CO_2. But in degassing we normally aim at a CO_2 residual of 5-10ppm, and the actual residual is not too critical because the object is to remove the bulk of the CO_2 for economy before its total removal by ion exchange. By contrast, the O_2 residual after deaerating is usually critical, and the levels required vary between 0.1 and 0.007ppm (the latter figure is equivalent to 0.005cc/l, which is the alternative unit for measuring dissolved O_2). Despite the lower solubility of O_2, deaeration is therefore a more rigorous problem.

Table 26.1. Henry's Law constant for O$_2$

Temp, °C	H to give x in mol/mol	H to give x in ppm O$_2$
0	2.58×10^4	1.45×10^{-2}
20	4.06×10^4	2.29×10^{-2}
40	5.42×10^4	3.05×10^{-2}
60	6.37×10^4	3.59×10^{-2}
80	6.95×10^4	3.91×10^{-2}
100	7.10×10^4	3.99×10^{-2}

$$x = \frac{p}{H}$$ where p is in bars and x is given in mol/mol or ppm O$_2$

Henry's Law and the Gas Law

Dissolved gas in equilibrium with a vapour obeys the formula

$$x = \frac{p}{H}$$

where x is the concentration of the gas in solution and p is its partial pressure (PP) in the vapour. H, the Henry's Law constant, is fixed for a given temperature.

The PP of a gas is its mole fraction in a vapour times the total pressure of the vapour.

If liquid and vapour are not in equilibrium, the gas diffuses in or out of the solution in an attempt to reach equilibrium. We therefore say that a gas in solution 'exerts' the PP with which it would be in equilibrium.

The volume taken up by gases at low pressure (which include water vapour) is fixed by the Gas Law

$$PV = wRT$$

where P = pressure, bar
$\quad\quad\quad V$ = volume, m^3
$\quad\quad\quad w$ = weight of gas, g mols*
$\quad\quad\quad T$ = absolute temperature, °K
$\quad\quad\quad$ R, the Gas Law constant
$\quad\quad\quad\quad$ = 83.14×10^{-6} for these units.

*The g mol is the weight in grams divided by the molecular weight.

Example:

Find the volume of 1 gm O$_2$ at 20°C and 2.29×10^{-4} bar

$$V = \frac{wRT}{P} = \frac{\frac{1}{32} \times (83.14 \times 10^{-6}) \times (293)}{(2.29 \times 10^{-4})}$$
$$= 3.324 \text{m}^3$$

A simple view of deaeration

Gases are removed from solution by bringing the water into contact with a vapour in which the partial pressure (PP) of the gas we want to bring out of solution is lower than the PP exerted by the solution. In degassing, a cheap source of such a vapour is air, which has a very low PP of CO_2. It can be used in large excess and freely returned to the atmosphere, carrying with it the CO_2 released by the water. In deaeration, things are more difficult.

The PP of O_2 in the atmosphere is about 0.2 bar, which is in equilibrium with about 10–14ppm O_2 in cold water, depending on the actual temperature, so we can't use that unless we apply vacuum to reduce the total pressure and with it the PP of O_2. Obviously vacuum operation is difficult, and always carries the risk of a leak pulling O_2 into the plant.

Alternatively we can pass another gas across the water to give us a flow of vapour which initially contains no O_2 at all. Then the PP of O_2 in the apparatus is due only to the O_2 which has come out of solution. This gas can be nitrogen, or if the water is at boiling point, it can be water vapour. Either can be made to flow in countercurrent to the water to provide a scrubbing and carrying vapour. By scrubbing we mean bringing a vapour with zero or low PP of O_2 into contact with the water to bring out the dissolved O_2, and by carrying we mean the task of taking away this O_2 to avoid its PP building up as more and more O_2 comes out of solution.

Clearly steam is the most suitable vapour in a boiler feed circuit, especially where we want to heat the feed to boiling point. The commonest deaerator is therefore the heater deaerator whose principle is shown schematically in Figure 26.1.

In Figure 26.1 water at a temperature below the boiling point at the operating pressure enters at the top, and steam at the bottom. The water entering at the top condenses most of the rising steam and is heated to boiling point. The boiling water then goes on falling through the steam. The apparatus is above atmospheric pressure and has a vent orifice at the top. This diagrammatic view shows that there are two different zones in a heater deaerator: a heating zone at the top in which most of the rising steam condenses, leaving a small amount blowing out of the vent and carrying with it the O_2 which has come out of solution; and a scrubbing zone below, in which the water flows in countercurrent against the whole steam flow.

These two zones handle different masses of steam and water, which makes the deaerator more complex than a single countercurrent scrubber (such as a

253

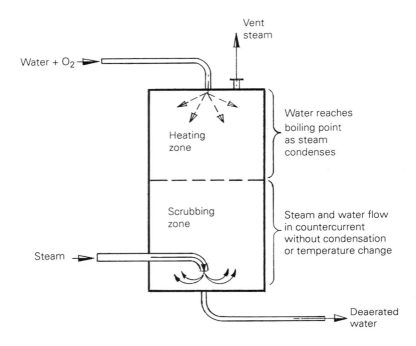

Figure 26.1. Heater deaerator

degasser). For following its theoretical basis it is probably simplest to consider first a flash vacuum deaerator, then a nitrogen scrubbing deaerator and finally the heater deaerator which is really our main concern.

The flash vacuum deaerator

For theoretical purposes let us consider a flash apparatus in which water is sprayed into a vacuum container, giving off dissolved gases and water vapour because of the low pressure; a vacuum pump maintains constant vacuum. There is no countercurrent flow in this kind of apparatus. Let us suppose we want an O_2 residual of 0.01ppm from water at 20°C. Taking $H = 2.29 \times 10^{-2}$ at 20°C, 0.01ppm O_2 would exert a PP of

$$P = (2.29 \times 10^{-2}) \times 0.01 \text{ bar} = 0.229 \times 10^{-3} \text{ bar} = 0.17 \text{mm Hg}$$

The vacuum pump must therefore get the PP of O_2 below this level.

The water gives off usually about twice as much N_2 as O_2, so that at equilibrium condition the PP of N_2 might be 0.46×10^{-3} bar. The saturated vapour pressure (SVP) of water at 20°C is 235.7×10^{-3} bar. Ignoring traces of helium etc, the total gas at equilibrium conditions consists of

$$0.23 \times 10^{-3} \text{ bar}$$
$$0.46 \times 10^{-3} \text{ bar}$$
$$235.7 \ \times 10^{-3} \text{ bar}$$
$$\overline{236.4 \ \times 10^{-3} \text{ bar}}$$

To get O_2 to come out of solution down to 0.01ppm we therefore have to get below this total pressure. As it is already within 0.69×10^{-3} bar of the SVP of water, we shall obviously end up with the water at boiling point and giving off the flash steam which is necessary to carry away the O_2.

The volume of vapour removed

The object of taking away vapour is to keep the PP of O_2 down, and therefore the calculation of the volume which has to be taken away is governed entirely by the mass of O_2 coming out of the water and the PP at which it is taken away. In the simple flash aparatus which we have used for this example, it must be taken away at 0.23×10^{-3} bar or less.

If we know what mass of O_2 we have to take away, we can work out the volume which that amount of O_2 occupies at that PP; this is the minimum volume to be extracted. The other constituents of the vapour will of course be contained within the same volume; they act as carriers and increase the total pressure at which the pump works.

As an example, suppose we are deaerating a feed water which is condensate and contains only 0.1ppm O_2 to start with. Then each tonne of feed contains 0.1gm O_2 of which we have to remove 0.09gm to obtain a residual of 0.01ppm O_2.

From the Gas Laws we can calculate that 1gm of O_2 at 20°C and 0.229×10^{-3} bar occupies $3.324m^3$, and from that we get 0.09gm O_2 at 20°C occupying $0.3m^3$ of vapour per tonne of feed. The total pressure of this vapour is 236.4×10^{-3} bar and the PP of O_2 in it is 0.23×10^{-3} bar. These are theoretical limits; in practice we have to extract more vapour at slightly lower pressure. It calls for some pretty expensive pumps.

As another example, if we feed cold make-up containing 10ppm O_2 then the mass of O_2 to be removed is roughly 100 times greater, and the volume to be removed is 100 times greater, still at the same high vacuum. This makes the pumps even more expensive.

Nitrogen scrubbing

The vacuum pump problem could be eased by increasing the total pressure with addition of more gas to serve as scrubber and carrier. By adding, say, N_2 into our flash apparatus the PP of N_2 can be increased from 0.34mm Hg to any level we like, depending on how much N_2 is added. This addition does not affect the PP of O_2 or the volume of vapour to be removed, but it increases the total pressure and eases the vacuum pump duty. If we add enough N_2 to bring the total pressure above atmospheric we can vent the vapour direct to atmosphere without a pump.

On the other hand, by raising the total pressure with N_2 we stop the water boiling, which reduces the rate at which O_2 will come out of solution. We shall need a more efficient apparatus with longer contact time to achieve the same result.

A more efficient way of using N_2 scrubbing is in a countercurrent apparatus, which works exactly like the scrubbing section of the heater deaerator shown in Figure 26.1. N_2 scrubbing is not common practice but it provides us with a good example for showing how the scrubbing section of a deaerator works without involving the complications which arise from the condensation of steam in a heater deaerator.

Suppose we have a countercurrent apparatus in which water enters at the

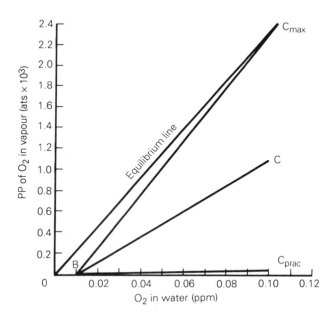

Figure 26.2. Concentration of O_2 in water/vapour

top and N_2 at the bottom. In countercurrent flow, the vapour then rises through water of ever-increasing O_2 content, so we can allow the PP of O_2 to build up as the vapour rises. We now get the same system of operation as degassing shown in Chapter 15. One difference in detail: Figure 15.1 of that chapter uses CO_2 concentration in air as one variable, which can be done for a process working only at atmospheric pressure. In deaeration, both pressure and concentration vary, so that in Figure 26.2 we must use the true variable which is the PP. Once again we have a straight line BC which describes the actual PP of O_2 in the rising vapour in relation to the PP of O_2 actually exerted by the water with which the vapour is in contact. Another straight line (OC), which is effectively a graph of Henry's Law, shows the conditions of gas and liquid in equilibrium. As long as the operating line BC lies below the equilibrium line, there is a driving force available to make the apparatus work, but of course the closer the lines are together, the smaller this driving force will be.

The slope of the operating line BC depends on the volume of N_2 flowing. If it is large, then the PP of O_2 in the vapour will not build up very much, and the line will keep low. The more N_2 we use, the lower the line BC and therefore the greater the driving force available to get O_2 out of solution. If we reduce the volume of N_2 flowing, the operating line BC will run steeper and steeper until we reach the limiting condition represented by BC_{max}, where C_{max} just touches the equilibrium line. This means that at the point where the vapour leaves the apparatus, in contact with incoming feed, vapour and liquid are just in equilibrium. This line represents the lowest flow of gas which is theoretically possible, and it calls for an infinitely efficient apparatus because at the point C_{max} the driving force is zero.

If we try to reduce the flow of vapour further, then on paper the operating line would cross the equilibrium line, as though O_2 were coming out of the vapour and being dissolved in the water.

The flow of vapour

We can easily calculate the minimum flow of vapour from the conditions at C_{max}, which represents a vapour in equilibrium with the incoming water. Let us take the conditions of our first example, where the incoming water contains 0.1ppm O_2 at 20°C. This has a PP of O_2 of 2.29×10^{-3} bar. From the Gas Law, as before, 0.09gm of O_2 removed per tonne then require a vapour flow of $0.3m^3$.

As a second example we took a make-up feed with 10ppm O_2 at 20°C, which exerts a PP of 0.229 bar. We remove from this almost 10gm O_2 per tonne of water, which at this PP occupy $0.033m^3$. These sums show that in countercurrent deaeration the theoretical volume flow is nearly constant for varying amounts of O_2 in the water; moreover the volume is much smaller than that required in a flash apparatus.

We can achieve the desired flow by using either a small mass flow of N_2 and working under vacuum or a large mass of N_2 at atmospheric pressure; in any case we have to use a good deal more than the theoretical minimum in order to lower the point C to a point where a reasonable driving force is available.

Nitrogen deaeration may actually be used in special circumstances. Some power stations start from cold every morning in order to meet daytime peak load. At start-up, the heater deaerators in these stations have no steam to work them, and for a short time some O_2 gets into the boiler circuit. When this happens daily, it can result in corrosion, and the operators have experimented with nitrogen deaerators to operate only during start-up. N_2 is expensive, but for the very short periods involved it might prove economical.

The heater deaerator

Feed water normally has to be heated anyway, and if it is heated to boiling point the heating system can provide an ample source of scrubbing vapour. Even a simple heater box, in which water cascades against a rising flow of steam, all of which is condensed, will remove most of the O_2 from make-up water. If the feed heater is designed specifically to be a heater-deaerator, then very low O_2 residuals are possible.

The heater deaerator as shown diagrammatically in Figure 26.1 is seen to have a heating and a scrubbing zone. Very roughly, the heating zone's function corresponds to a heater box and the scrubbing zone's to the N_2 deaerator described above. In the scrubbing zone the actual numerical relationships are different because of the high temperature at which heater deaerators operate, which changes the value of the Henry's Law constant and increases the temperature term in the Gas Law calculation. For example, at 100°C, H = 3.99×10^{-2} and 0.09gm of O_2 at a PP of 0.1ppm in solution can be calculated to occupy 0.022m^3. This compares with the calculation (above) for conditions at 20°C, when the corresponding volume was 0.03m^3, so the difference in the theoretical minimum vapour flow is not very great. This is because the rising value of H with temperature gives us higher partial pressures, but the rising temperature then works in the opposite direction to give us larger volumes.

Many people say that deaeration at high temperature is easier because the solubility of O_2 is lower. This is not really true, because in fact H does not change all that much. The real differences are that in the heater deaerator the water can easily be contacted with a very large flow of vapour, and at higher temperatures the rate at which O_2 comes out of solution is much greater.

In fact we can hardly help using a vast excess of steam. For control purposes, heater deaerators are rarely operated with less than 10°C temperature rise and, in a well-designed deaerator, most dissolved O_2 comes out in the heating zone. Suppose we run a deaerator at 110°C, feed it with water at 100°C, and assume that by the time the water arrives at the scrubbing zone it contains 0.1ppm O_2.

The 10°C temperature rise needs 18.8kg heating steam per tonne of feed, which in these conditions occupies 22.3m^3, 1000 times the minimum we have just worked out as theoretically necessary to carry away 0.1ppm O_2. With a greater temperature rise, the excess will be even greater.

Figure 26.2 shows that this steam flow gives us an operating line BC_{prac} which is almost horizontal and therefore provides the maximum possible driving force.

Carrier steam

This ample flow of steam all condenses in the heating zone so we have to arrange for a little bit more to pass right through the heater deaerator in order to carry the O_2 through the heating zone and out through the vent. A good design reduces the vent steam to the lowest possible value in order to conserve heat and water.

For example, one commercial design specifies only 150kg/h vent steam at a throughput of 1200t/h, that is 0.125kg steam/tonne, or 0.0125 per cent. If, as in the above examples, we remove 0.09gm O_2/tonne, then the carrier steam will contain a mole fraction of

$$\frac{0.09}{125} \times \frac{18}{32} = 0.41 \times 10^{-3} \text{mol } O_2/\text{mol } H_2O$$

At our operating pressure of 1.4 bar the PP of O_2 in the carrier steam is then

$$0.41 \times 10^{-3} \times 1.4 = 0.57 \times 10^{-3} \text{ bar}$$

Taking H as 3.99×10^{-2}, this PP would be exerted by 0.016ppm O_2 in water, which compares with 0.1ppm O_2 which is actually coming in with the water at the vapour outlet point. This means that although we have concentrated enormously the O_2 in the scrubbing steam, there is still a sensible driving force tending to bring O_2 out of solution in the heating zone.

Similar arithmetic for the case of a 10ppm O_2 feed in similar conditions produces a vent steam in equilibrium with 1.6ppm O_2 in solution, so theoretically the relationship between vent steam and incoming vapour is unchanged at all incoming O_2 concentrations.

In practice, however, the vent steam and the heating zone represent the most critical problems of heater deaerator design. In the last example for a 10ppm O_2 feed, we ended up with 4.5 per cent O_2 in the vent steam, in equilibrium with 1.6ppm O_2 in solution. Theoretically this is only in contact with feed at 10ppm O_2, which is fine. But if by accident some of this O_2-laden vapour were to get back to the other end of the apparatus, then we would indeed be in trouble.

The heating zone

We worked out 18.8kg steam per tonne of feed for a 10°C temperature rise, to which must be added vent steam at, say, 0.125kg/tonne, which is all that is left passing through the heating zone. The incoming steam flow of 18.925kg is therefore reduced to 0.125kg, a factor of 151:1. Often the temperature rise is higher, and then the difference is even greater.

Steam condenses rapidly, so that the change in steam flow is abrupt. In the scrubbing zone the steam velocity might be 10m/sec; in the heating zone this brisk flow suddenly drops to a near-stagnant fog of O_2-laden vapour. The rush of steam towards the water surface may trap some incondensible gas, or slow-flowing vapour can be trapped by fast-falling liquid and carried in the wrong direction. Imperfect distribution of steam flow across the heating zone can cause the formation of stagnant pockets in which very high O_2 concentrations can build up. Any of these events can cause trouble.

A deaerator fed with condensate which contains little O_2 does not handle enough mass of O_2 to give real problems. But where the feed is make-up, which can contain up to 14ppm O_2, then the vent steam O_2 can approach atmospheric concentrations. One way to combat trouble then is to increase the vent rate to reduce the O_2 concentration in the vapour and to increase vapour velocities in the heating zone. This results in the costly nuisance of a great plume of vent steam blowing to atmosphere. The proper solution is to design the deaerator with care, so that these problems do not arise.

Vacuum deaerators

The heater deaerator combines two tasks in a simple and satisfactory process, but sometimes it is undesirable to heat the feed above 100°C at the deaeration stage and the deaerator must operate under vacuum. Any leakage, however, pulls atmospheric O_2 into the plant and defeats the object of the exercise, especially where low residuals are wanted. Most vacuum deaerators are therefore simple devices which do not aim at a low residual, but merely reduce the O_2 for economy before chemical scavenging.

Summary

Deaeration is a much more rigorous problem than degassing because of the very low O_2 residuals which are usually wanted. The best and commonest solution is the heater deaerator, in which the flow of feed heating steam supplies the large volume of vapour which is needed to achieve very low partial pressures of O_2. When this heating steam condenses, there remains a very small flow of carrier steam highly enriched in O_2. With a badly designed apparatus this can affect the quality of the product.

27 The practical design of heater deaerators

To sum up the theoretical rules governing the design of heater deaerators: a deaerator which can take significant amounts of O_2 out of water to give very low O_2 residuals consists essentially of two zones. One zone heats the water, the other scrubs out the last traces of O_2.

The scrubbing zone can only take the O_2 down to a low residual if there is a large flow of steam through it. This carries away the O_2 released by the water and dilutes it so much that the partial pressure build-up of O_2 is very small. Such a flow of steam can in effect only take place past water which is at the boiling point.

The heating zone serves to heat the water to the boiling point. Provided the water is heated through a sufficient temperature range, then the demand for heating steam will automatically provide far more steam than is theoretically wanted in the scrubbing zone. Most of this large flow condenses in the heating zone, leaving only a very small amount of carrier steam, highly enriched in O_2, which it vents to the atmosphere. The design must take special care to prevent this enriched vapour from somehow contaminating the scrubbing zone.

Besides these theoretical necessities, there are practical ones such as achieving the necessary mass transfer, of controlling steam and water flow and of avoiding re-contamination of the deaerated water.

Choosing the pressure

The boiler cycle as a whole dictates the pressure at which the deaerator has to work; the arrangement of heaters, economizers and condensate returns usually leaves little choice. Theoretically a deaerator can operate at any pressure, provided the flow sheet allows for enough difference between feed and deaerator outlet temperature to provide the necessary temperature rise. In practice, vacuum deaerators should be avoided if possible as they are both costly and troublesome, but sometimes there is no alternative.

At the other end of the scale, the cost of the pressure vessel discourages very high pressures, especially where the deaerator is combined with a large storage tank. Usually the pressure is below 10 bars, and for medium pressure boilers it is more likely to be around 2 bars. Vent steam from the deaerator is normally controlled by an orifice. For control purposes it is best to operate at pressures above 1.15 bar (104°C).

Storage capacity and location

Somewhere within any deaerator the water is allowed to break pressure and cascade freely. Every design must therefore include some storage for collecting the water and feeding a pump suction, usually the boiler feed pump. In most installations this storage is made big enough to serve as the hot well in the system. This means that the majority of deaerators are associated with a large tank.

Some deaerators are therefore designed to sit on top of the tank; sometimes designers put two or more deaerator domes on a single tank. Other deaerator designs are intended to go inside the tank which excludes them from the minority of projects where no storage is associated with the deaerator. Those deaerators which are not combined with storage are usually vertical cylinders; the deaerated water is collected in a small volume at the bottom of the tank which holds just enough for control and then feeds the pump.

The water coming from a deaerator is at boiling point at the operating pressure. To provide enough Net Positive Suction Head for the pump which follows, the deaerator has to be mounted high. Sometimes the siting of the deaerator dictates the necessary building height.

Achieving mass transfer

Chapter 26 showed that quite a small heating range gives a flow of steam through the scrubbing zone which completely swamps the tiny amount of O_2 released in this zone. For all practical purposes the partial pressure of O_2 throughout this zone is then nil and it therefore doesn't matter whether the flow within the zone is countercurrent or not.

The driving force available for bringing O_2 out of solution is therefore as high as it can possibly be, but it is still extremely small. This means that we need a large area of steam–water interface, or a long contact time, or both.

Practical reasons make packed towers unsuitable for pressure deaerators: none of the cheap materials of which packings can be made is fully suitable, nor do packings readily accommodate the sudden changes in steam and water flow which must be expected. On the other hand, packed towers are normally used for vacuum deaerators, whose lower temperatures allow plastic packings to be used and in which the conditions do not change so drastically.

Plate 16. Prefabricated deaerator on its way to a major chemical complex

Steam supply to the deaerator is usually controlled to sustain the pressure in the vessel. At start-up and load change, therefore, the plant may 'gulp' steam; unless carefully controlled this can cause the violent changes mentioned above. Any baffles, trays or other obstructions must be carefully designed to withstand the resulting stresses. There are numerous examples where they have buckled or ripped out. Some deaerators 'bang' or 'rattle' under certain conditions and, in time, corrosion or fatigue may lead to mechanical breakdown.

Another way to avoid these dangers is to have no trays or baffles in the steam path and extend the steam–water interface area in some other way. Rather than install metal surfaces on which the water spreads out, designers use sprays of water in steam or bubbles of steam in water.

263

Achieving countercurrrent flow

Sprays cannot give perfect countercurrent flow, and steam bubbling through water even less so. But as we saw above, countercurrent flow within the scrubbing zone is not important. The important point is to avoid the concentrated vapours of the heating zone getting into the scrubbing zone, that is to achieve perfect countercurrent flow between the two zones. The best way is to have some physical separation between the zones.

This requirement can be relaxed only where the deaerator duty is not demanding, either because the feed is already low in O_2 or because the O_2 is not critical. A few examples of deaerator designs will show how these principles are put into practice.

The heater box

This is a crude and very cheap deaerator whose object is to heat the water to just below 100°C while taking out the bulk of the O_2. It therefore represents the heating zone of a proper pressure deaerator. As Figure 27.1 shows, it is an open baffled tank in which water cascades in countercurrent against steam. Steam flow is controlled by the outlet temperature and theoretically there is no vent steam at all. Nevertheless the O_2 is reduced to 1ppm or less.

The spray-and-tray deaerator

This kind of deaerator can be used with or without a storage tank. Figure 27.2 shows it in the form of a dome to be mounted on a tank; by itself it would extend a few feet downward on the cylinder to provide the necessary storage with the outlet going straight to the pump section.

Water is sprayed in at the top of the dome and the sprays form the heating zone. By the time the water reaches the top tray it is just about boiling. It then cascades over perforated doughnut and plain trays which serve to hold up the water flow and re-divide it, in order to provide the time and interface area needed to remove traces of O_2.

This design and many like it have given good service for years but they have their limitations. Even with good design, the sprays and trays are points of potential weakness, especially on the large scale.

Neither of the designs shown in Figure 27.1 and 27.2 has a positive separation of the heating and scrubbing zones. The heater box is of course not meant to give low O_2 residuals. By the same token, the spray-and-tray deaerator cannot always give low residuals if the feed contains a high level of O_2. It will, however, give good service, for example in electricity generating stations where the feed is almost all deaerated condensate and contains less than 0.1ppm O_2.

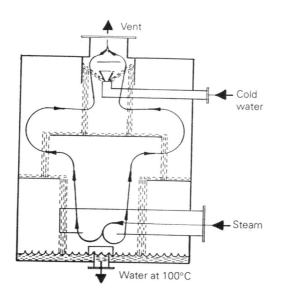

Figure 27.1. Heater box

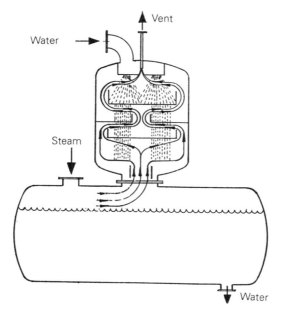

Figure 27.2. Spray-and-tray deaerator

Plate 17. A complete water-treatment plant installed in Cuba (note the high-level deaerator)

A scrubber can be added to the spray-and-tray design to give the separate section necessary to scrub out the last traces of O_2 under all feed conditions. Figure 27.3 shows such a scrubber incorporated into the storage tank. Water falls out of the dome into the tundish and is scrubbed by steam bubbling through it.

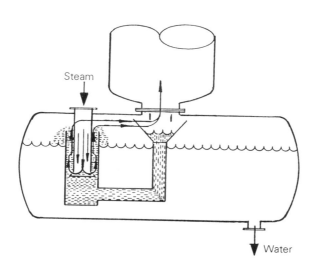

Figure 27.3.   Scrubber design

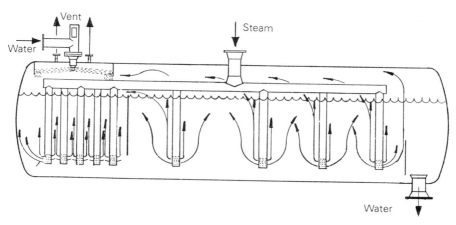

Figure 27.4.   Stork-type deaerator

267

An actual example from the field demonstrates these effects rather well. A spray-and-tray deaerator as shown was first commissioned to produce a residual of less than 0.005ppm O_2 from a deaerated condensate, and gave satisfactory service for quite a time. Then the feed changed to un-deaerated make-up, and from this increased O_2 load it proved impossible to get the specified residual. A satisfactory residual was obtained by boring out the vent orifice plates to increase the vent steam more than tenfold, but the wasteful nuisance of a huge steam plume could not be allowed to continue indefinitely. A scrubber was fitted which allowed the vent to be reduced to an acceptable level while still obtaining the desired output quality.

We have seen that increase in vent flow only made a small difference to the scrubbing zone steam flow. Its effect must have operated in the heating zone, where it probably swept away and diluted the rich O_2 vapour. With a separate scrubber this became unnecessary.

The Stork-type deaerator

This is one of several designs meant to go inside the storage tank, which need a tank holding at least 10 minutes' worth of water at full flow. One great advantage of building the deaerator in the tank is that the unit can be assembled on the ground and then lifted into place. Assembly can even include insulation and access platforms. As deaerators usually go high up under the roof, this is quite important. Another point is that a tank without a dome can give several feet more NPSH within the same ceiling height.

The Stork deaerator has a spray chamber serving as a heating zone, and the rest of the whole storage tank operates as a bubbling scrubber as shown in Figure 27.4. As there are no baffles restricting the steam path, the chances of serious mechanical damage are slight. With careful design, this kind of deaerator will give the specified O_2 residual over a wide range of flow rates with any incoming O_2 concentrations. The internal fittings are relatively expensive if they have to be mounted in a small tank but this kind of design becomes progressively more attractive as the size of tank increases.

Part 10
POLLUTION AND EFFLUENT TREATMENT

28 General

Diminishing water resources and increasing public awareness of pollution over the last three decades have resulted in the installation of large numbers and many types of waste-water treatment plant. Here we outline methods of liquid treatment in common use and broadly explore effluent categories and polluting effects.

Control of pollution

Pollution is difficult to define precisely, as may be observed from legislative instruments which in certain cases employ the words poisonous, noxious and polluting in relevant sections. Statutory instruments in the UK require that discharges of sewage and trade effluents (with certain notable exceptions) have the consent of the National Rivers Authority (River Purification Board in Scotland). It therefore falls to these statutory bodies to translate the legislation into practical control measures. The authority may refuse consent, grant consent unconditionally or grant consent subject to conditions or standards which should take account of the quality of the water required in the receiving watercourse. In addition to safeguarding water quality as such, the standards must also ensure the welfare of the naturally occurring animal and plant life of the stream. To achieve these aims, the conditions applied to limit polluting effect may by physical, chemical and/or biological parameters.

271

Classification of polluting effect

Pollution may be categorized by the impact which an effluent type exerts on the water quality and ecology of the recipient. Industrial waste waters may be discharged directly to a stream or, in urban areas, to the sewerage system. The standards applied in the former case will be more stringent than in the latter. For sewer discharge, applied standards must ensure that the objectionable constituents are at a level acceptable to the sewage disposal works, the effluents from which are also subject to 'consent procedure'.

A variety of sources and processes produces polluting substances which may be broadly classified into physical, inorganic, organic and bacterial contamination. The effect of a single constituent may often be predicted, but combinations of substances or types may create difficulty in control as a result of synergistic effects.

Physical pollution

Effects of this type of pollution are in the main indirect. Examples are temperature increase, colour and solids. Temperature rise may be caused by natural sunlight and warm trade effluents, the most obvious of which is water used for cooling purposes. Disturbance of both the biological and chemical regimes occurs in the receiving watercourse and the solubility of gases decreases. In terms of gas solubility, dissolved oxygen concentration is critical to most types of animal and plant life, and depletion to well below saturation level can have far-reaching effects. Rise in temperature also causes an increase in animal metabolism which combined with other adverse factors may result in distress or death in extreme cases.

Colour may also give rise to problems, although the coloured constituent need not exhibit a toxic or other polluting effect. The colour may act as a barrier to sunlight, thus affecting plant growth normally brought about by photosynthesis. Fish life may be indirectly affected by the reduction in food crops and by an inability to detect food sources.

Suspended matter, even if inert, is undesirable, resulting in siltation in sluggish stream sections, blanketing of plant life and, in extreme cases, clogging of the gills of fish. The effects may be observed even when the suspended solid does not demonstrate a toxic, chemical or biological reaction.

Inorganic pollution

This may result from inorganic substances which react with the plant and animal life, or with the stream water itself to alter a normally satisfactory regime.

Toxins are the most objectionable and noticeable of these substances, and the appearance of dead fish is disturbing to the general public. Toxins, depending on their character and physiological effect, may produce instant mortalities, or certain poisons may accumulate to critical level over an extended period. The best known and most common are the heavy metals such as zinc, copper, chromium, lead, cadmium, nickel. Cyanides, halogens, ammonia and common acids and alkalis may also fall into this category and exhibit toxicity if present in sufficient concentrations.

Apparently harmless salts and constituents may create adverse stream conditions, depending on the environment into which they are discharged. Acid solutions of ferrous salts reaching an alkaline stream will produce an unsightly precipitate of ferric hydroxide, which will also clog and restrict the growth of plant life. Excess nitrogen present in the form of ammonia, nitrite and/or nitrate, together with a source of carbon and phosphorus, can upset the ecological stream balances and result in eutrophication. This latter phenomenon may be particularly objectionable when the recipient is a lake or estuary where displacement of the water containing the relevant constituents is limited by the topography of the area. Common salts may be regarded as pollutants by virtue of their ability to create a change in environment and ecology below a point of discharge.

Organic pollution

This is probably the type of pollution most commonly encountered. While not always detectable visually, organic matter is present in domestic sewage and a large number of liquid trade wastes. Regardless of source, most organic material is of animal or vegetable origin, and includes starches, sugars, organic acids, proteins and other nitrogenous compounds.

These natural constituents may, in the case of certain trade wastes be present singly in small or trace amounts, and may not by identifiable or measurable. Under most circumstances, a number of these compounds is present, and the concentration of such matter is expressed empirically as a cumulative oxygen demand.

Oxygen demand is important in the control of pollution as it is essential to maintain aerobic conditions in stream water and in the biological treatment sections of effluent treatment facilities. Both biochemical oxygen demand (BOD) and chemical oxygen demand (COD) are commonly used parameters (the nomenclature is derived from the nature of the tests involved) and give an indication, as a single convenient figure, of the additive effect of the organic impurities listed above.

Other organic pollutants of note are toxic componds such as phenols, cresols and others derived from trade processes, together with pesticides such as aldrins and chlorinated hydrocarbons. These latter are particularly objectionable to the environment, and compounds such as polychlorinated biphenyls may accumulate in animal cell tissue, resulting in long-term effects on food chains.

Bacterial pollution

Bacterial contamination usually originates from animal sources, including humans, domestic animals or birds. The most serious pollution of this nature usually occurs downstream of a sewer or sewage works discharge, since in most other cases dilution factors obviate any likely nuisance. The exception to this general rule applies when a carrier of a specific waterborne disease resides in an area (possibly sparsely populated) where the sewage receives a very poor standard of treatment. If there is free access to the receiving watercourse, a danger of spread of disease exists. However, in the UK cases of acute bacterial contamination are negligible, and vigilance is only necessary when abstraction from the recipient for potable use is contemplated.

Plate 18.   Ash-handling system under construction at a coal-fired power-station

Plate 19. Dairy effluent treatment plant under construction at Lord Rayleigh's dairy

Methods of treatment

Physical methods

Cooling may be achieved by means of ponds, cooling towers and heat exchangers. Ponds are simple but inefficient, and the most common system used is the evaporative cooling tower. Natural draught towers are used for power-station cooling water duties, while the more efficient fan-assisted variety is commonly used by industry when recirculation of the cooling water is practised. Fairly sophisticated packings may be employed to ensure efficient cooling. Water must therefore be maintained in good condition to prevent build-up on the packing of solids in the form of inorganic scale or biological film. Heat exchangers are used for indirect cooling but these units are normally used as process tools to conserve heat. Heat is transferred into the cooling medium which will often be used in production processes.

Screening is used for removal of large suspended solids. A great variety of screening equipment exists, ranging from open bars to fine mesh or wedge wire having 1mm apertures. Cleaning methods vary greatly from manual raking to mechanically driven rakes or brushes. Alternatively rotating screens with 'doctor knives' and backwash sprays may be used. These devices can usually be subjected to automatic operation.

The proper selection of the screen will depend on the particle size and nature of the solids to be removed. The screening area required will depend on the flow rate and the screenings load to be removed. Screens are seldom (except in the case of storm sewage) a means of treatment in themselves. They are normally employed as pretreatment to reduce easily removed impurities and/or for the protection of pumps, tanks and other integral parts of the complete treatment system. Screens are suitable for removing paper, rags, strings, faecal matter, vegetable peelings, animal gut, grain and numerous other materials present in waste waters.

Gravity separation is commonly used for removal of solid material. While sand and grit may be removed in small sumps and specially designed and profiled channels, large settling tanks are usually employed. Such tanks may operate on the fill-and-draw principle, or as continuously fed units of varying geometry. The most common are of rectangular 'horizontal' flow and circular radial and upward flow patterns. The design objective is to induce suspended matter to settle within the tank, removing not only suspended solids but, in the case of organic wastes, a proportion of the organic load associated with the suspended matter.

Tank sizing depends largely on the settling velocity of the solids, and broadly the velocity of flow through the tanks should be appreciably less than the solids' settling velocity. A great deal of experimental and research work has been carried out on settling tank design, and numerous formulae have been advanced incorporating sophisticated design parameters. However, empirical approaches are still used in the main for design, with parameters such as surface overflow rate and retention time.

The principles of sedimentation, as set out in Chapter 3, apply equally here. In this context, however, we are dealing with particles which are both bigger and heavier than alum floc, so that the apparatus is simpler. Its successful operation depends largely on the ease and efficiency with which the settled solids can be removed from it.

Settlement or sedimentation tanks are employed mainly as an integral unit operation within a more complete treatment system, although in the handling of storm sewage and a few inorganic wastes, settlement alone may give a satisfactory degree of treatment. In the case of organic wastes, sedimentation processes alone are unlikely to produce an effluent suitable for discharge to a recipient watercourse unless an extremely large dilution is available.

In sewage treatment, sedimentation is used in a pretreatment role to reduce solids and thus organic load on the biological systems. Worthwhile solids/ organic load reductions may also be achieved on meat processing, tannery, certain vegetable processing effluents and paper wastes.

It will be observed from the above that gravity separation removes suspended material which has a greater specific gravity (SG) than water. The solids settle into the tank base and form a wet sludge. Gravitational forces may be used in open tanks to separate material with an SG less than water. In this case, the materials float to the surface and the sludge/float may be skimmed off. The principle is particularly applicable to the separation of oils, fats and grease. Other immiscible organic materials have been handled in this way. Various devices have been employed within sedimentation tanks to assist gravitational forces to achieve the desired effects. Slow-speed stirrers and fixed baffles, together with sludge blanket development, have been used to promote collision, and consequently coagulation of small particles. In addition, fine particles may be removed by application of a tilted plate separator pack installed in purpose-designed tanks. The system is applicable to the separation of particles from a carrier liquid where the differences in gravity are low and is in common use for the removal of mineral oils from numerous effluent types.

Assisted separation   is here defined as the development of forces to augment natural gravity for achieving separation. Hydrocyclones and centrifuges are typical of types of plant used, together with systems of the dissolved air, froth and electro-flotation types.

Hydrocyclones effect separation of particles by means of centrifugal forces developed in the units themselves. Solids are thrown to the peripheral wall of the unit and the only energy used is the kinetic energy naturally present in the waste-water flow. The system is suitable mainly for the removal of such heavy particles as sand and grit.

Centrifuges appear in different forms; that in most common use in the effluent treatment industry is of the horizontal solid bowl design. Centrifuges function by means of mechanically developed centrifugal forces within the units, the prime mover being an electric motor. The main application is for sludge dewatering, although specific trade waste waters may benefit from pretreatment by these machines for removal of solids and starches.

Flotation processes (see Chapter 4) cause separation of solid particles from waste water by means of gas bubbles. Bubbles are generated in, or introduced to, the effluent in a flotation chamber by various means and adhere to the solids present. The rise of the bubbles through the liquid carries the solids to the surface in the form of a sludge float which can be skimmed off for disposal or re-use. Power, in one form or another, is required for bubble generation and the system may on occasion by used as a sole treatment. Flotation methods are particularly suitable for removal of oils, fats, greases and recovery of protein; other uses include sludge thickening and final separation of solids generated in biological treatment systems.

Chemical treatment

Chemicals may be employed for destruction of undesirable materials, sterilization and pH control. Perhaps the main use is in the precipitation/coagulation of polluting constituents in solution or suspension for subsequent removal by one or other of the separation processes previously outlined. Examples of chemical destruction are the oxidation of phenols with hydrogen peroxide, and chlorination of cyanides.

Several compounds are in fairly common use for disinfection (see Chapter 8) including ozone, chlorine and chlorine dioxide. The materials, when added to produce an appropriate residual concentration, kill pathogens. It is employed normally as a final treatment since other polluting material may produce an excessive demand for the sterilizing chemical.

Precipitation of undesirable components from an effluent may be achieved by the addition of a reagent to combine with the constituent to produce an insoluble salt. This is removed by a physical treatment method, such as sedimentation or flotation.

In the main, this type of treatment has found favour in the handling of inorganic wastes, but with advances in the preparation and application of synthetic polymeric materials for the improvement of floc formation, interest in these systems has increased over the last decade. Mechanical equipment for flotation processes has improved, resulting in a wider range of applications for the techniques. Organic wastes are now being handled satisfactorily and, indeed, each food and drink waste is usually scrutinized for potential value as a protein source in an overall examination of disposal methods. An interesting chemical treatment used increasingly in Europe is precipitation for removal of phosphate from treated sewage effluent to prevent the occurrence of eutrophication in the recipient.

Biological treatment

This is the most common method used for dealing with organic waste waters. It is employed where extra fine, colloidal and dissolved organics cannot be removed satisfactorily by physical or chemical treatment. Waste waters having organic material of this nature are numerous and include, among others, sewage, food and drink, coke-oven and paper-making effluents. Bacteria and fungi use the impurities as food and energy sources and bacterial metabolism converts the organic material into cell tissue, carbon dioxide and other, less harmful, simple by-products. The insoluble material formed as cell growth is used to continue the purification, while the normal excess is removed regularly as sludge. This, and the sedimentation or flotation required to achieve it, is an integral part of the system.

Two main methods are employed for biological treatment of organic wastes:

Activated sludge is an aerobic treatment system functioning by cultivating a floc in the waste water being treated. The floc, which must be kept in suspension in the liquid waste, consists of a range of biological life forms. Impurities are adsorbed and absorbed by the activated sludge as the two are continuously brought into contact by mixing devices in large tanks. These tanks are known as aeration bays, and an adequate supply of oxygen is essential to satisfy the oxygen demand of the impurity and the respiration rate of the sludge floc. The oxygen supply and mixing is usually effected by the same device, being introduced by the entrainment of air by the mixer or aerator. In some cases, however, a separate air supply is provided, and the most recent developments of the system involve the use of pure oxygen.

The aeration tanks are designed to provide a suitable period of contact (almost invariably on a continuous flow basis), and the mixture of waste water and activated sludge solids flows to a sedimentation tank where the activated sludge is separated. The sludge solids are returned to the aeration tank at a rate sufficient to maintain an adequate concentration of active organisms for continuation of purification of the waste water, while the clarified liquid overflows the tank weir as treated effluent. Activated sludge in excess of the requirements for purification is diverted to a sludge-handling system.

A number of variations of the activated sludge principle have been developed, but these fall mainly into the high rate, conventional and extended aeration classification. These may be categorized by organic load (or oxygen-demand load) of the waste water applied per unit volume of the aeration tank. The main types of system may be sub-divided further by the oxygenation and mixing method. Methods used include mechanical mixers on either horizontal or vertical shafts which introduce air or oxygen into the waste water at the liquid surface; alternatively, air or oxygen may be introduced by means of diffusers, sparge pipes or venturi tubes.

Biological filtration systems (see Chapter 29) are the oldest biological systems, having developed from land treatment (sewage farm) techniques. Filtration in this context is a misnomer since, as with the activated sludge system, purification is carried out by bacteria and no physical filtration takes place. The main distinguishing feature of this method is the state of mobility of the bacteria. In the activated sludge system the bacterial growths live suspended in the waste water while in the filtration system the bacteria are provided with suitable surfaces on which to grow. Hence the name fixed-film reactors now sometimes used to describe the process. Air and oxygen supply is by natural ventilation.

The most common method employing fixed films is the bacteria bed. In this system, waste waters are evenly spread over the plan area of the bed by some form of distribution device. The liquid percolates by gravity through the bed, which is packed with media to provide the surface on which the organisms grow. A mixture of purified effluent and voided bacteria emerges at the bottom of the bed, and the dead and decaying organisms are removed by settlement. The system differs further from the activated sludge process in that there is no need to recycle the settled solid which is removed regularly from the settlement tanks as a sludge for disposal.

Many variations of the biofiltration process exist. The traditional 'low rate' bacteria bed employs minerals as the support media. The specific surface area-to-void ratio of such materials is, however, limiting and, although improvements were achieved with the introduction of recirculation and double filtration techniques, a major advance followed the introduction of plastic packings. These materials can be moulded easily to give optimum void to surface area, thus minimizing the risk of blockage in use. A new flexibility in the use and design of biofiltration plants became possible, this being particularly relevant to the development of high-rate/high-loaded systems.

The most recent development of the fixed-film process was the introduction of the 'rotating biological contactor'. These units operate by providing, to the film growth, surfaces which are rotated while partially submerged in a tank of waste water. The organisms come into contact regularly with the impurities and are then exposed to the air for supply of oxygen requirements. Variations arise from the type and configuration of the materials supplying the surface area for bacteria. These include rotating discs, tubes and baskets containing randomly disposed units of plastic media.

General

The above unit processes are well established and improved variations have been, and continue to be, introduced. Although the basic technology is dated, the systems described can either singly or in combination meet the normal standards applied by the National Rivers Authority and HM Inspectorate of Pollution, and, indeed, the demands of the recipients themselves.

More sophisticated treatment may be employed to augment the traditional processes where very high-quality effluents are required, among others: physical filtration (through sand or similar granular material), activated carbon adsorption, ion exchange and membrane separation techniques. These are in the main used when dilution factors in the receiving body of water are negligible, and/or water re-use is contemplated.

Plate 20. Aerial photograph of industrial waste-treatment plant

Of biological treatment techniques, only aerobic processes have been covered in this chapter. Anaerobic processes have been used in the past mainly for the digestion of organic sludges. Recently, however, a great deal of interest

281

has been generated in the use of anaerobic treatment for partial degradation of strong organic wastes and Chapter 30 covers the subject. At the other extreme, such processes may be applied to nitrate-rich water or good quality effluents for the removal of nitrate by reduction to nitrogen gas.

The extent and method of treatment selected must depend, in large part, on the quality of the effluent to be achieved. This is normally determined by the standards set by the National Rivers Authority. Proper river-basin management will ensure that the uses of the river are not impaired and that it may continue as a source of process, potable, irrigation or cooling water and for recreational purposes, navigation or simply as an acceptable effluent carrier.

29 Biological filtration

Introduction

The introduction to a watercourse of biodegradable organic matter, which consumes dissolved oxygen and promotes growth of microorganisms, leads to water pollution. If oxygen is reintroduced as with naturally flowing water, by diffusion from the atmosphere, the organic matter and deleterious microorganisms will be destroyed and oxygen in the water will again be available for beneficial marine growth.

The trickling or percolating filter is a recognized device for rapid aeration of domestic or industrial waste containing biodegradable organics. The heart of the filter is the medium or packing. The medium serves as the support system on which the slime of microorganisms grows. The waste percolates by gravity over the slime-covered medium. Air rises through the medium by natural draught, usually countercurrent, to the downflowing waste, and the aerobic microorganisms in the slime digest the organic matter in the presence of oxygen which diffuses into the slime.

Percolating or trickling filters are in use all over the world for the aerobic biological purification of sewage and organic waste waters. They were first used on a large scale in 1893 at the Salford, UK, works and have been proved to be a highly dependable oxidation process.

Filter installations, compared with the activated sludge process, are more tolerant of continual or shock discharges of certain toxic industrial wastes (for example, containing heavy metal ions, phenol, cyanide, formaldehyde, sulphide). In addition, acidic or nitrogen- and phosphorous-deficient wastes from the food and drink industries, commonly coincident with hydraulic overloading, are tolerated better.

Over 66 per cent of the municipal plants in Western Europe and the USA use this type of system. In a survey of 422 UK sewage works, serving populations greater than 10 000, 30 per cent of the total flow was treated by filtration alone in 286 plants, 48 per cent by activated sludge alone in 97 plants, and the remainder in works employing both processes together.

Biological filtration systems can be broadly categorized into low, intermediate and high rate, depending on the hydraulic or organic loading applied to the medium. It is generally agreed that sharp demarcations do not exist between the successive categories and that to some extent the descriptions are interpreted differently according to local practice. It is however becoming universally accepted that high-rate biofiltration systems start at approximately $3m^3/m^3$/day hydraulic loads or with organic loads in excess of 0.6kg BOD/m^3/day. Low-rate biological filter systems usually employ random mineral media, starting at $0.5m^3/m^3$/day and 0.1kg BOD/m^3/day respectively.

Low- and intermediate-rate filters are usually employed to remove low weights of BOD at high efficiencies (90 per cent plus removal at organic loads of 0.1kg/m^3/day) and high-rate filters to remove large weights of BOD at medium efficiencies (60 per cent removal at organic loads of the order of 3.0kg BOD/m^3/day).

Low-rate biological filters are usually circular or rectangular beds of graded mineral media, usually in a concrete or brick containing-structure. In the past the media used were clinker, slag, stone, gravel, coke and in some cases coal (notably at Bradford) and were usually packed to a total depth of 1.8m in England and Wales. However, in Scotland, whinstone medium is quite often used for low-rate filters and packed depths up to 3.6m are common. In other European countries local practices at considerable variance with these general guidelines are common. British Standard BS1438 covers the most important properties of mineral media which are defined as:

- freedom from dust;
- durability;
- resistance to chemical attack;
- surface area;
- weight and density;
- surface roughness;
- size distribution.

High-rate biological filters are most often built above ground and are either circular or rectangular in plan area. They normally contain plastic media and packed depths can be much greater due to the high voidage and low material weight. The introduction of plastics has eliminated a number of the problems associated with mineral media:

- there is normally no dust;
- plastics are durable and resistant to chemical attack;
- they have uniform size distribution; and
- they can be manufactured to meet a specific surface-area requirement.

Principles of biological filtration

Low- and intermediate-rate systems

Purification of water-borne waste containing organic components by treatment on low-rate biofilters is a biological process which relies on absorption of both soluble and suspended matter from the waste water into and on to local slimes which develop and proliferate on the surfaces of the filtering medium over which the water trickles or flows, followed by further complicated processes of decomposition and synthesis of both the soluble substances and sludge-forming solids. Some of the solids are thereby removed from the wastes. These decomposition and oxidation processes result from the activity of bacteria, various forms of fungi and protozoa, together with other living organics which feed upon them, preventing unrestrained accumulation and clogging.

When wastes are fed continuously and with only short interruptions on to such filter beds, a covering of slime develops gradually during a maturing period of from several weeks to six months. After this period, full purification of the wastes proceeds continuously, providing aerobic conditions are maintained within the bed. The effective zone of the slimes for aerobic biological activity exists only down to 2–3mm depth below the slime surface because deeper zones become septic. The degree of purification is dependent on the effective surface area of filter medium available for direct contact with the waste. The total slime surface, and hence the active surface per unit volume of filters, depends on the shape and size of the filling material.

One of the most important features of a low-rate biological filter in terms of purification efficiency is retention time. This is particularly important as the reactions involved are comparatively slow, particularly nitrification. The processes of purification in a low-rate filter can essentially be split into three parts:

- coagulation stage — coagulation and flocculation of colloids and pseudo-colloids;
- oxidation stage — oxidation of carbonaceous matter to carbon dioxide;
- nitrification stage — oxidation of ammonia, derived from the breakdown of nitrogenous organic matter to nitrite and eventually nitrate.

With increased loading and lower residence time within the filter the degree of nitrification will fall.

Percolation of water down through the bed also results in some flushing effects. In low-rate filters, the rate of application of waste waters is low enough to ensure substantial decomposition of the organic matter within the filter. Portions of active as well as decomposed or dead inactive slime material are progressively washed down through the bed with the percolating water and finally appear as a flocculant suspension known universally as humus. Discharge of humus solids from filters goes on all the time, but the amount discharged may vary widely from time to time. Filters which are in good condition remain unaffected by this continuous loss of slime.

The amount of solids finally discharged is relatively low. However, larger amounts of accumulated slime are unloaded from the filter at intervals of many months (perhaps 2–4 times yearly), such periods lasting for some weeks. The humus solids from such filters tend to be granular. In low-rate filters, ammoniacal and nitrogenous organic compounds are usually well enough oxidized to yield high proportions of oxidized nitrogen, principally in the form of nitrates with some nitrite.

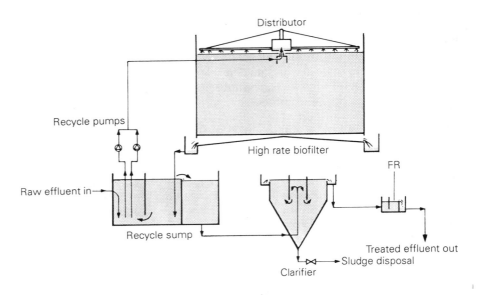

Figure 29.1. Typical biofilter treatment system

High-rate systems

High-rate filtration did not become recognized as a viable method until the advent of plastic media which became available commercially in 1963. Prior to this time, some early work on a limited scale had been carried out using coarse mineral media, notably at Dunstable (UK). There are now over 2000 plants in operation on a worldwide basis using plastic media in the role of high-rate filtration.

One of the main advantages of plastic packings over conventional biological filtration media is that they give a better performance; they have a higher work capacity per unit volume, that is higher rates of BOD are removed per unit volume of medium. Full use is made of this advantage in the design of high-rate biofiltration units.

As the majority of high-rate filters are built with medium depth between 3 and 8 metres, taking full advantage of the light-weight nature of the system, the effluent invariably requires to be lifted to the top of the filter by pumping, compared with many gravity-motivated mineral low-rate filters. Even so, compared with the operating costs of some other well-tried effluent treatment systems, considerable savings are found which can be as high as 80 per cent. Effluent flows down the packing and biomass in a thin film. Because of the uniform structure a supply of air is allowed to rise through the filter and a supply of oxygen is transferred through the falling film of effluent to the biomass. This enables the process of BOD reduction to take place at high rates.

Similarly, as with conventional percolating filters and the activated sludge process, the following humus settlement tanks form an integral part of the treatment process. The effluent is collected from the base of the filter and then passed to a gravity settlement system. One effect of plastic media treatment, in this form, as distinct from bio-oxidation effects, is to make it possible to separate by sedimentation a greater proportion of the suspended or colloidal material present, and to employ higher flow rates in sedimentation vessels. Solid voided from these systems settles very readily and upward flow velocities of 2.4m/h are possible.

Experience on full-scale plants of treating domestic sewage has shown that with an intelligent approach to the desludging of settlement tanks, a sludge of 3–4 per cent dry weight solids can normally be removed. The sludges are readily amenable to treatment by anaerobic digestion processes.

Initially it was believed widely that the success of high-rate biofilters in carrying out partial treatment was due to the preferential biodegradation of a supposedly easily treatable fraction. No evidence, either in the field or experimental, has emerged to show that the remaining fraction is less reactive.

Engineering considerations

All biological filters (both low and high rate) consist of the same basic parts:

- medium
- support and drainage system

- filter enclosure
- distribution system
- ventilation provision.

Media — low-rate filters

Historically, crushed rock was the most commonly used medium, although lavas showed considerable advantage when available, having lower effective density. Mineral media with a rough surface tend to mature more quickly and are generally more suitable for biofilters.

Theoretically smaller pieces of medium should be superior to large pieces because the surface area is greater the smaller the size. Mineral media of 40–80mm effective diameter expose approximately $100m^2/m^3$ of surface. A size range of 25–40mm exposes approximately $200m^2$ and, as a consequence, filters filled with 25–40mm rock might expect to purify approximately twice as much sewage as those with the coarser material. However, although the total volume for water, air and slime is approximately the same in each case, the dimensions of the interstices are very much smaller for the finer rock and this effectively limits hydraulic loads, ventilation and the relative quantity of slime growing. Because of this, improvements in efficiency of only up to 50 per cent for a tenfold increase in surface area of mineral media have been experienced.

However, the introduction of plastics media has overcome this problem. A random plastics medium with a surface area of $330m^2/m^3$ has been tested in comparison with mineral media of $100m^2/m^3$ and has demonstrated an efficiency increase of 400 per cent over the mineral media.

In England it was customary to fill beds with rock of about 35–50mm, laying 100–150mm rock at the bottom for about 0.3 of a metre, sometimes using a special topping layer of 15–25mm at the surface. The recommendation in Germany is for filling of about 40–80mm uniformly throughout the bed, and similarly in the USA a maximum of 60–100mm size has been found most satisfactory. Thus a uniformly distributed filling of 40–80mm size, with possibly a coarse layer at the bottom, appears to be the best for most cases. For shallow filters, a maximum size of 55mm is suggested.

The depth of the filter bed must be limited to suit the size of medium, having regard to natural ventilation, unless forced ventilation is provided. In England, beds of 1.4–2.7m in depth are common, a typical arrangement being 1.8m depth of rock of about 40mm. In Germany depths of 1.8–4.3m with media of 50–75mm give satisfactory results, and in the USA, with 64mm rock beds of 1–2.5m depth are used.

Random plastic media are made in uniform sizes and allow beds to be filled with the same material throughout their packed depth. These media normally offer a higher surface area per unit volume than do mineral media.

Media — high-rate systems

As one would expect, the desirable characteristics of a high-rate medium differ from those of a low-rate medium. A high-rate medium should:

- be capable of removing high weights of BOD per unit packed volume;
- be capable of operating at high hydraulic loads per unit volume and unit area;
- possess a significantly open structure to avoid blockage by solids and to ensure an adequate supply of oxygen without recourse to forced aeration;
- be sufficiently strong, structurally, to bear its own weight and the weight of overlying layers of medium, together with the attached growth of biomass;
- be sufficiently light in weight (even when loaded with biomass) to enable a significant reduction to be made in the civil engineering costs of plant construction;
- be biologically inert, neither attacked by nor inhibiting the growth of the treatment biomass;
- be chemically stable, not degrading with use or in the presence of small quantities of solvents, organic chemicals etc; and
- have as low as, or cheaper cost per kg of BOD removed when packed as conventional biological purification processes.

Plastic media are usually used to meet these requirements, and the surface area and voidage ratios can be optimized to suit particular applications.

Plastic media configuration falls into two categories, either random or ordered structure. Random media are usually in the form of injection moulded rings as cylindrical collars with internal and external projections to increase surface area. The rings are simply poured into a retaining tank to form a bed of uniform thickness. Due to the random placing, localized bridging and channelling of liquid flow may occur. Surface area ranges from $100–300m^2/m^3$ and material of manufacture is usually polypropylene.

Ordered plastic media are supplied in modules used as building blocks. Modules are constructed from sheets of moulded plastic joined on their vertical faces to form a rigid box structure with vertical channelling. This material can be stacked as a self-supporting mass within an enclosure formed by lightweight cladding, on a wood or steel sub-frame. Ordered media provide a very open structure for the passage of air and sloughing solids but require a relatively high liquor flow per unit surface area of filter. Surface area is normally $90–100m^2/m^3$ but some examples provide up to $230m^2/m^3$. Material of manufacture is usually PVC.

Support and drainage system

Media can be supported off the base slab by the use of intermediate-bed or coarse material, or purpose-designed drainage tiles, to allow free drainage of liquor to central or peripheral collection zones, as well as the free passage of ventilating air.

Alternatively, the medium can be supported on a grid floor above the base slab, which provides greater capacity in terms of ventilation and drainage. The system adopted will depend on process considerations, biofilter size and density of the medium.

Filter enclosure

As previously mentioned, the enclosure may be a simple frame and cladding system for self-supporting ordered plastic media or in the form of a containment tank for random media and traditional mineral-medium filters.

The usual configuration adopted is circular, and traditionally low-rate filters were of concrete construction. With deep-bed plastic media, circular bolted- or welded-section coated steel shells are utilized for random media, and straight-sided treated timber frames with trapezoid profile lightweight sheeting for ordered structure media. In all systems, adequate openings at the base must be provided for free ventilation.

Distribution system

Circular filter shells permit the use of rotary distributors which provide the most effective system, and this has been the traditional method. High-rate systems demand a multiple arm arrangement to handle the high irrigation rates required. Alternatively, a fixed spray system is installed as part of a piped distribution network, normally adopted for square or rectangular filter configurations only.

Ventilation provision

In most filters natural ventilation follows from differences in density between the atmosphere and the air inside, which may result in the movement of air either upwards or downwards in the filter. Moisture and carbon dioxide content may significantly affect the relative density of the air but temperature is the principal factor. The temperature of the filter air approximates to that of the percolating water, and in temperate climates is usually colder in summer and warmer in winter relative to the atmosphere. The direction of ventilation is therefore mainly downwards in summer and upwards in winter. However, especially in summer periods, the direction of ventilation may change twice daily, the waste-water and filter-bed temperatures remaining practically constant while the atmospheric temperature commonly varies as much as 6–12°C (or more) above and below the water temperature.

Experimental data show that air temperature differences of 6°C cause natural ventilation at the rate of $18m^3/m^2/hr$. Taking the filter dosed with sewage at a loading rate of 0.9m/hr, a temperature difference of 6°C would provide ventilation at the rate of $20m^3/m^3$ sewage. Since somewhat less than $1m^3/m^3$ of sewage would provide sufficient oxygen, an oxygen absorption efficiency of only 5 per cent is adequate for filtration. This indicates that even relatively small temperature differences, say only 1°C, may generally be sufficient for adequate ventilation. Furthermore it has been found that wind velocities exceeding 0.9m per second cause significant ventilation. Artificial ventilation is sometimes used for small packaged units and in particular situations. For example, filters which are totally enclosed, because of cold climate or to combat odour or fly nuisance, must be continuously ventilated mechanically.

Systems design

Volumetric loading rates are commonly expressed in terms of the total volume of waste applied daily per unit volume of filter medium over which the waste is applied (m^3/m^3 day). In the UK, for domestic sewage using mineral media, the average design rate used for single filtration is $0.6m^3/m^3$ day, based on a flow of 180 litres per head per day. Expressed as organic load, this is equivalent to 0.1kg BOD/m^3 day. When random plastic media are employed the flow figure increases to the order of $2m^3/m^3$ per medium day or 0.3–0.4kg BOD/m^3 day.

Recirculation of effluent may be helpful in the filtration process. Treatment of strong wastes or relatively strong sewages may be more successful if the strength is lowered by dilution. Normally, this is achieved by recirculating the filter effluent. This is the most effective method since some biological activity, including enzymes with oxygen and nitrates, is then introduced into the influent waste. This refreshes the waste and acts as a buffer against acidity, reduces odours and generally supports subsequent aerobic purification. A portion of the effluent is normally returned after sedimentation and mixed with the settled waste before it is distributed on the filter bed. However, where a random plastic medium is used, it is normal to recycle unsettled filter effluent thereby minimizing the cost of subsequent humus settlement. When recirculation is employed, the load figures proposed for single filtration can be substantially improved upon. There is, however, an economic limit to the amount of recirculation because of pumping costs and the larger final settling tanks required in low-rate systems. High recirculation ratios are frequently necessary when dealing with very strong organic trade effluents.

Other successful variations of trickling filtration have been developed. With the arrangement of two filter beds in series, in the process known as two-stage filtration, the effect is much the same as for a filter with a depth equal to the sum of the primary and secondary filters, but there can be some advantage with two relatively shallow filters because they are more readily ventilated, also because additional aeration may result from the secondary distribution of the primary effluent.

Alternating double filtration is another type of two-stage filtration. Each filter may be provided with its own settling humus tanks, and piping and channels are connected with pumps and so arranged that one of each of a pair of filters may be used alternately as primary and secondary filter, reversing the sequence regularly from day to day or otherwise as found expedient or desirable. This technique is applied to low-rate filtration and permits appreciable increases (say twice normal rate) in overall loading rates with the same degree of purification including nitrification. Alternating double filtration has been used successfully, especially in the treatment of dairy wastes where single filters soon become choked with fatty material and fungal slime.

High-rate filters are normally used for partial treatment and are not normally designed for the production of high-quality effluents. They are found usually in combination with low-rate aerobic systems (both low-rate filters and activated sludge systems) to produce high-quality effluents. The combination of unit operations with each operating at optimum efficiency produces the lowest capital-cost scheme.

Anaerobic systems

Recent developments have enabled the basic fixed-film biological filter system to be operated anaerobically, utilizing plastic media. Anaerobic digestion of pollutants has the attraction of possible utilization of a by-product, methane gas, as a fuel source.

Exclusion of air can be achieved by operating the bed in a submerged condition or simply by enclosing the filter totally. Unlike traditional batch-type complete mix digesters, the necessity for maintaining the contents in a homogeneous condition by injection of off-take gas or by mechanical agitation does not arise.

To be economically viable, anaerobic systems usually require a feedstock in excess of 10 000mg/l BOD. As the reaction takes place at elevated temperature, the cost of heating must form part of the economic evaluation.

30   Anaerobic digestion

Introduction

In the use of microorganisms for the removal of polluting substances, as part of treatment technology, the broad division is between oxygen-rich or oxygen-deficient environments known respectively as aerobic and anaerobic digestion. This technology essentially adopts the natural processes of the ecosystem and provides optimum conditions to accelerate reaction rates and control the process.

Anaerobic digestion involves the breakdown of almost all kinds of organic matter by the concerted action of a wide range of microorganisms (principally bacteria) in the absence of oxygen or other strong oxidizing chemicals. Methane and carbon dioxide are the principal end products, with minor quantities of nitrogen, hydrogen, ammonia and hydrogen sulphide (usually less than 1 per cent of the total gas volume) also generated. The bacteria responsible for this process are found, albeit usually in small numbers, in a wide variety of places: soil, lake sediments and animal intestines.

It is just over 100 years since these organisms were harnessed by man to hasten the breakdown of pollutants under controlled conditions.

Chemical pathways

Different kinds of organic wastes and waste waters may be used as raw material for methane production. The biodegradable components are converted to methane according to the simplified scheme shown in Figure 30.1:

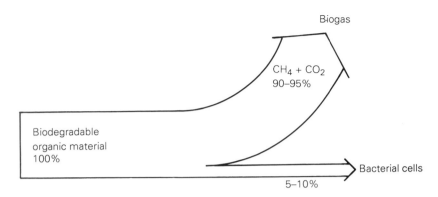

Figure 30.1. Waste conversion

Organic material is fermented by bacteria stepwise to methane (50–70 per cent, v/v) and carbon dioxide (50–30 per cent) with concomitant production of bacterial cells. The mixture of methane and carbon dioxide is called biogas and has an energy content of $18–24 MJ/m^3$, compared with a value of $37.3 MJ/m^3$ for pure methane. A considerable amount of the energy present in the raw material can thereby be recovered as methane.

A more detailed presentation of methane formation is shown in Figure 30.2. Complex organic material is hydrolysed to soluble organics and these are converted to volatile fatty acids. The higher volatile fatty acids, for example propionate and butyrate, are degraded to acetate, carbon dioxide and hydrogen and these compounds are used by the methane-producing bacteria as carbon and energy sources for the formation of methane.

The methane-producing bacteria are slow-growing and strictly anaerobic. In waste-water treatment systems they are completely dependent on other bacteria for the removal of oxygen from their environment.

Process design

The predominant application is for the digestion of sewage solids and farmyard wastes.

Conventional digesters (Figure 30.3a) are commonly used in developing countries for the anaerobic digestion of manure. The digester contains a mixture of manure and water. A certain amount of manure and water is added to the reactor daily and the same volume of reactor content is removed. The biogas produced is collected at the top of the reactor and stored in a tank. These represent the simplest form of digester and require the minimal capital outlay.

A major problem is to distribute the incoming flow of waste water evenly across a reactor bottom surface of maybe 50–100 m^2. Compared with the anaerobic filter the higher recycle rate imposes greater running costs.

No single type of reactor is suitable for every situation. Depending on the properties of the waste or waste water being treated, the most suitable process must be chosen.

Operational parameters

The following conditions must be met for anaerobic digestion:

- a waste water rich in biodegradable organic substance;
- a good nutritional balance (for example, BOD, N, P, micronutrients);
- absence of toxic compounds (for example, heavy metals, chlorinated hydrocarbons);
- pH6.8–8.0;
- temperature 30–60°C; and
- water content > 80 per cent.

Frequently measures have to be taken to correct an improper nutritional balance or to adjust pH.

Some substrates suitable for anaerobic digestion have been listed in Table 30.1. Many kinds of organic material can be digested. One notable exception is lignin, which is resistant to microbial attack under anaerobic conditions. Consequently, some waste waters from the paper and pulp industries are not well suited to methane fermentation without pretreatment to increase biodegradability. This pretreatment could be chemical or enzymatic.

The production of methane and its utilization as an energy source is an important factor in the consideration of treatment economics. In many applications it is used as the process technology for the production of methane fuel from purpose-grown crops or readily available by-products from food-processing operations.

The upper theoretical limit for methane formation is 350 l/kg COD. In practice, the methane yield is lower. Representative data are compiled also in Table 30.1.

Key factors for floc-based digesters

Mean biomass densities in floc-based digesters range from 5–15g/l for contact processes with low-strength wastes to 50–70g/l for sludge blanket reactors with

Table 30.1. Methane production achieved from different substrates

Substrate	Process	Methane production	Methane production per volume digester	Author
Municipal sludge	High-rate digester	$0.31m^3$/kg VS added	$0.68m^3/m^3$.d	Data from Käppala sewage-treatment plant Lidingö, Sweden
Municipal refuse	High-rate digester	$0.46m^3$/kg VS destr.	$0.59m^3/m^3$.d	Swartzbaugh *et al.* (1977)
Manure:				
beef	High-rate digester	$0.28m^3$/kg VS added	$4.5m^3/m^3$.d	Varel *et al.* (1977)
beef	High-rate digester	$0.35m^3$/kg VS added	$3.6m^3/m^3$.d	Hashimoto *et al.* (1977)
dairy	High-rate digester	$0.22m^3$/kg VS added	$2.4m^3/m^3$.d	Bryant *et al.* (1976)
dairy	High-rate digester	$0.24m^3$/kg VS added	—	Morris (1976)
Waste waters from food and fermentation industries:				
wine distillery	Contact process	$0.69m^3$/kg COD added	$1.9m^3/m^3$.d	Ross (1980)
guar gum	Filter process	$0.22m^3$/kg COD added	$2.3m^3/m^3$.d	Witt *et al.* (1979)
beet molasses	Fluidized bed	$0.15m^3$/kg COD added	$3.3m^3/m^3$.d	Frostell (1980)
sugar beet	UASB process	$0.30m^3$/kg COD destr.	$3.9m^3/m^3$.d	van der Meer *et al.*(1980)
Special crops:				
water hyacinth	High-rate digester	$0.23m^3$/kg dry weight	—	Wolverton & McDonald (1979)
giant kelp	High-rate digester	$0.17m^3$/kg VS added	$0.48m^3/m^3$.d	Klass *et al.* (1979)

Notes: VS = volatile solids   COD = chemical oxygen demand

medium and high-strength wastes. Irrespective of whether floc/liquid separation occurs within the digester or in a separate vessel (tank or centrifuge) high biomass densities require high floc-settling velocities, higher than the liquid velocity at the point of separation. This in turn requires large flocs, heavy flocs and low liquid velocities.

Floc size

While biological aggregate formation is still poorly understood, the formation of large flocs appears to require:

- Conditions of low turbulence. This applies in the case of the UASB and filter processes, where gentle mixing is caused by rising gas bubbles. Smaller flocs seem to be found generally in contact processes in which mechanical mixing is used.
- Particular bacterial growth conditions. A low food:microorganism ratio has been found to favour floc formation. This is achieved by maintaining low organic loadings during digester start-up.
- Use of a flocculating agent. Synthetic polymers have been shown to increase floc size and settleability.
- Selection pressure. The use of upflow biochemical reactors has been shown to favour the retention of flocculated microorganisms.
- Other factors. Other suggested influences on floc formation are the presence of high levels of certain cations (for example Ca, Ba) and the absence of certain kinds of finely divided matter.

Floc density

Floc density is increased by:

- Minimizing the formation of filamentous forms of bacteria (bulking sludge) by operating with an adequate supply of growth nutrients.
- The inclusion in flocs of inorganic precipitates. This will depend on the presence of heavy metals in the digester feed.
- Minimizing the entrapment of gas bubbles in flocs. This is achieved by:
    - minimizing the rate of gas production at the location or time of floc settling by: plug flow of waste through the digester, with little waste remaining and hence a low rate of gassing near the liquid outlet (for example UASB); a cyclic feeding pattern, with little waste remaining at the end of the feed cycle; rapid cooling of flocs, reducing biological activity; aeration of flocs, inhibiting gas formation;
    - removing gas bubbles from flocs. Methods of achieving this are:

301

agitation caused by the rising gas bubbles themselves (UASB and tower); mechanical agitation (contact process); the presence of solid packing-materials which cause rising flocs to follow a tortuous, turbulent path, for example stones/Raschig rings, balls, mesh as used in anaerobic filters and the contact process; the application of a vacuum to digester effluent.

Liquid velocity

Having maximized floc-settling velocities, upflow liquid velocities in the direction of the liquid outlet have to be kept sufficiently low to enable flocs to settle. This is achieved with a settler and baffle design which directs rising gas away from the settling zone and provides sufficient cross-section for the velocity required.

Key factors in film-based digesters

While the performance of fixed-film and fluidized-bed reactors has been well documented for laboratory and pilot-scale operations, little information appears to be available to allow an economic appraisal of each type. However, an analysis of fluidized-bed reactors for aerobic BOD removal showed that the most significant costs were for oxygen supply and recycle pumping, while the support medium (sand) and the reactor vessel itself were not significant cost components. The conclusion which can be drawn from this for an anaerobic system (with no oxygen requirement) is that attention should be focused on minimizing the recycle flow in order to reduce overall process cost. This can be achieved by several means:

- The extent of bed expansion. Fluidization should be minimized, without producing unfluidized, bypassed pockets; 30–100 per cent bed expansion is typical.
- The biomass film support particles should be of low density and small in order to minimize their fluidizing velocity. There is some motivation for finding cheap, robust support materials of lower density than silica. Alumina and resins, ceramics and activated carbon have been used although support costs have not apparently been reported.
- Expanded or fluidized-bed processes become more attractive where waste strength is low. At the short waste-retention times which result, liquid recycle is minimized, with much of the bed fluidization being brought about by the raw waste throughput alone.

Scale-up of fluidized-bed digesters mostly involves an increase in bed cross-section in order to maintain constant upflow velocities. However, as the bed cross-section increases it becomes more difficult to distribute the fluidizing liquid uniformly. At least one company claims that it has overcome this problem.

Improved biomass activity in digesters

In parallel with developments in digester design which have increased biomass concentrations in anaerobic systems, microbiological and biochemical advances have led to increases in digester biomass activity, and improvements in the stability of the digestion process.

Advances in biochemical understanding

Perhaps the most important single advance is the understanding of the role of hydrogen gas in regulating methane formation. Hydrogen is formed in digesters during the fermentation of wastes containing carbohydrates, fats or proteins and is removed by the formation of either methane or acetic acid.

It is now appreciated that, under high-substrate loading conditions or after a sudden increase in substrate load, volatile fatty acids and hydrogen can be formed more rapidly than they are removed. A key group of organisms (the obligate hydrogen-producing acetogens) which exist in a close symbiotic association with methanogenic bacteria are inhibited if the hydrogen concentration increases above 0.01 per cent (in the case of those organisms converting propionate to acetate, CO_2 and hydrogen). Higher fatty acids and hydrogen continue to accumulate, the digester pH drops and ultimately methane production ceases. This type of process failure may be the mechanism underlying many digester breakdowns in the past.

A practical application of this understanding is the possibility of using hydrogen analysis (in addition to pH and individual fatty-acid analyses) to monitor digesters. Gas chromatographs designed specially for analysis of low levels of hydrogen are available commercially. If automated, this analysis could give an early warning of impending instability.

Digestion with phase separation

As the two broad groups of digester bacteria (acid-forming and methane-forming species) differ widely with respect to their growth rates and optimal growth conditions, there has long been interest in separating the acid- and methane-forming populations in two serial reactors. This was pursued throughout the 1970s at the Institute of Gas Technology in Chicago. Recently, workers at the University of Amsterdam have investigated this approach in considerable detail and have obtained results of great practical value.

Complete acidification of a glucose-based substrate has been achieved at extremely high space loads under optimal pH (5.8) and temperature (30–37°C) conditions. Using an upflow sludge-blanket process for this work, very low biomass yields were obtained (2–7%), thus maximizing the yield of acids available for methane formation and minimizing problems of sludge disposal.

In carefully controlled comparisons of one- and two-phase digestion of soluble substrates it was found that:

- Biomass activity was increased. The maximum biomass activity of the slow (methane-forming) step of the two-phase process was 1.5kg COD/kg VSS/day compared to 0.5kg/kg/day for a conventional, single-phase system. This three-fold improvement has been calculated to increase the maximum space loading of a digester with a biomass content of 30g VSS/l from 12–13 kg COD/m³/day to 40–45kg COD/m³/day.
- Digester stability was improved. The two-phase approach eliminates digestion instability caused by the accumulation of hydrogen, propionic and butyric acids following shock load increases.

Improved digester stability and biomass activity following presouring of carbohydrate wastes has been confirmed in practice. Waste presouring (phase separation) is incorporated in sludge-blanket processes engineered and marketed by the Institute for Gas Technology (USA) and Studiebureau O. de Konickx (Belgium).

Bacterial nutrition

Increasingly, it is being realized that it is not adequate simply to ensure that the levels of nitrogen and phosphorous are sufficient for stable digestion. Initial work on the nutrient requirements for bacteria digesting a wide range of wastes was reported in 1964. The elements N, P, S, K, Mg and Fe were found to be required, with trace levels of metals as found in tap water. No amino acids or vitamins were required.

Since then the role of some trace metals has become known. Cobalt is essential for the formation of methylcobalamin, a coenzyme involved in methane formation. Nickel has been found to be the central metal in a nickel tetrapyrrole called F430. This is a key part of another enzyme essential to the formation of methane from H_2 and CO_2. Similarly, low levels of molybdenum are required in digesters. A requirement for iron, originally indicated by Speece and McCarty, has recently been demonstrated with a range of wastes.

The roles of metal precipitation and the formulation of soluble metal complexes in digesters have been presented, with implications for the nutritional availability of metals for bacterial uptake in digesters.

While much of this work on nutritional aspects of digestion is at an early stage, any new waste for which anaerobic digestion is proposed should be thoroughly tested for all known nutrients, using empirical testing procedures available in the literature.

31   Sewage treatment

Introduction

Domestic sewage is a mixture of waste waters arising from household waste facilities including toilets, bathrooms and kitchens. It contains pathogenic microorganisms (bacteria and viruses that cause disease) and creates odour nuisance (sulphurous smell). Most sewage-treatment plants are designed to purify domestic sewage to an effluent quality adequate for discharge to the nearest watercourse. An exception exists in arid climates where waste waters are often used for landscape irrigation and other industrial and agricultural purposes.

Sewage is typically about 99.9 per cent water; the other 0.1 per cent is mainly organic materials (organic means derived from living things), a wide range of compounds comprising the elements carbon, hydrogen and oxygen plus smaller amounts of nitrogen and sulphur. The organic impurities enter the water during its various household uses.

The majority of the organic material is biodegradable, that is it can be consumed by microorganisms (bacteria etc.) and thereby removed from the water. The materials are converted to more microorganisms (which appear as sludge), carbon dioxide (which is a gas that escapes to atmosphere) and soluble salts including nitrate and sulphate. The high biodegradability of the organic material allows biological processes to be used to purify the sewage.

Treatment parameters

A typical chemical analysis for raw sewage is shown in Table 31.1. For the UK the Royal Commission Standard of 30mg/l solids and 20mg/l BOD has been the established target for discharge to watercourses. However this standard was based on a minimum dilution by the receiving waters of 8:1. Increasingly with ever higher abstraction and water usage many watercourses cannot provide this dilution. In addition, especially in arid countries where water is scarce, re-use of the treated effluent is required. Both situations demand more stringent levels of treatment. Thus primary and secondary stages are being augmented by tertiary treatment to achieve levels of 10mg/l solids and 10mg/l BOD or better. The most common laboratory test used for sewage is the BOD_5 test (the five-day biochemical oxygen demand at 21°C). This test gives a measure of the concentration of biodegradable organic material in the sewage or effluent and is effected by measurements of the dissolved oxygen removed from the water by the microorganisms while feeding on the organics. A sewage sample is diluted with oxygenated water and the dissolved oxygen concentration measured. The sample is placed in an incubator (temperature-controlled container) at 21°C for 5 days. At the end of this period the oxygen concentration is measured again and subtracted from the original reading to give the amount consumed by the microorganisms. This is then adjusted by the dilution factor to give the BOD_5.

Table 31.1. Average (24h composite) analysis of crude sewage

Analysis	Concentration (mg/l)
COD	650
BOD	326
Organic carbon	173
SS	127
Organic N	19
Ammoniacal N plus urea nitrogen	47
Total nitrogen	66
pH (value)	7.8
Anionic detergents	16
Sodium	100
Potassium	20
Calcium	110
Magnesium	7
Chloride	70

Ammonia (NH₃) is also present in the sewage as a result of natural breakdown of urea (urine). Ammonia is not organic because it does not contain carbon. It is not reduced significantly by the same microorganisms that remove organic materials. It is removed mainly by nitrifying microorganisms (e.g. nitrobacter). This process is called nitrification and results in the conversion of ammonia to nitrate.

It is important when ammonia removal is required that for nitrification to be complete it is necessary for the sewage to contain at least 8mg/l alkalinity (as CaCO₃) for every mg/l ammonia removed (as N). This can be a problem with sewage derived from desalinated waters. In such designs plant modification may be necessary (addition of alkalinity and/or denitrification).

Plate 21. Rotating biological contactor being installed at remote housing complex

Another common laboratory test used for sewage is the suspended solids test (SS). This test measures the concentration of undissolved solids present and is effected by filtering the sample and weighing the solids deposited on the filter paper.

Daily variations in sewage flow

Fortunately for the designers of sewage-treatment plants, sewage characteristics are fairly consistent world-wide. Unfortunately the pattern of everyday life introduces large variations in sewage flow over the 24 hours of a day. Older combined foul and surface water collection systems will show peak levels during high rainfall. An example of household water-flow variation is given in Figure 31.1.

A community comprising mainly facilities, with husbands going to work, children going to school and most wives staying at home, does have a fairly predictable sewage-flow pattern. In this case most of the sewage will be discharged during 16 hours per day with very little flow during the 8 hours during the nights. The flow will peak during the morning, after lunch-time and in the early evening. At these three peak periods the flow could typically be 3 times the average daily flow, each time for about one hour.

In contrast to this typical domestic community is, for example, the all-male workers camp. In an extreme case all of the daily sewage could arrive during 2 two-hour periods which are early morning and early evening at a flow of perhaps 6 times the average daily flow.

It follows that the larger the community served by a treatment plant the greater will be the attenuation of localized peak flows and the ratio of peak:average flow to be treated will be reduced.

Thus, without collection and flow balancing, all treatment plant carries the burden of catering for wide flow variations in its hydraulic engineering. Fortunately the active biomass utilized in most treatment plants has the ability to absorb pollutants rapidly on contact and the controlling parameter is not hydraulic residence time. Typical values for sewage flows are given in Table 31.2.

Sewage treatment processes

There are many processes that can be used to purify sewage:

- *physical processes* include screening, settlement, flotation, sand filtration, membrane filtration, reverse osmosis and distillation;
- *chemical processes* include coagulation and flocculation using

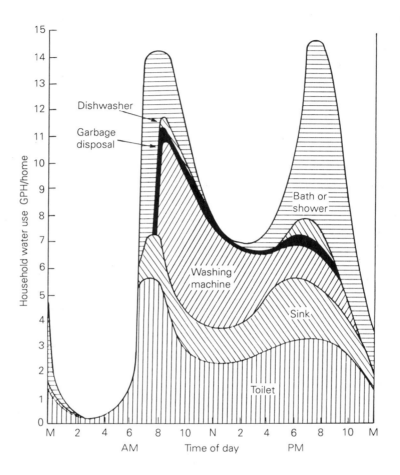

Figure 31.1.   Daily household water use

polyelectrolytes, lime, iron salts (ferric chloride or ferrous sulphate) or aluminium salts (aluminium sulphate), oxidation using chlorine, hypochlorites, ozone or peroxides and adsorption on activated carbon;

• *biological processes*

  • where microorganisms are freely suspended in the water, such as activated-sludge processes including extended aeration and contact stabilization

  • where microorganisms are attached to a surface, such as biological (trickling, percolating) filters or filter beds, and rotating biological contactors (biodisc, biosurf, biospiral etc).

Table 31.2. Guidelines for the average flow of sewage from various establishments

Type of establishment	Volume of sewage (l/head/day)
Small domestic housing	120
Luxury domestic housing	200
Hotels with private baths	150
Restaurant (toilet and kitchen wastes per customer)	30–40
Camping site with central bathhouse	80–120
Camping site with limited facilities	50–80
Day schools with meals service	50–60
Boarding schools — term time	150–200
Offices — day work	40–60
Factories — per 8 hour shift	40–80

All of these processes have been used at some time, either in a full-scale plant or on an experimental basis and Figure 31.2 shows the layout of a large treatment plant. From the results of plant operation, experiments and the associated capital costs and running costs certain economical combinations of processes have become preferred for the purification of domestic sewage:

Preliminary treatment Simple physical processes at the plant inlet to remove materials that may damage, block or inhibit the main treatment processes. These are usually screening (to remove debris) or comminution (to shred debris), often sand and grit removal by flotation and sometimes flow equalization.

Primary treatment Removal of settleable solids from screened sewage usually by settlement but occasionally by very fine screening or flotation. Primary treatment produces an odorous and infectious sludge (primary sludge) which usually requires further treatment. For this reason it is often omitted and the downstream processes modified to suit the additional solids load (as in extended aeration).

Secondary treatment Removal of organic materials from screened or settled sewage by biological processes. Aerobic microorganisms are much more effective than anaerobic in the proportion and speed of BOD removal. Also, aerobic conditions assist in the removal of pathogenic microorganisms. Aerobic processes require oxygen to be dissolved in the water. This can be achieved by the injection of pure oxygen but it is more common and economic to use mechanical (surface) aerators or diffused (bubble) aeration to transfer

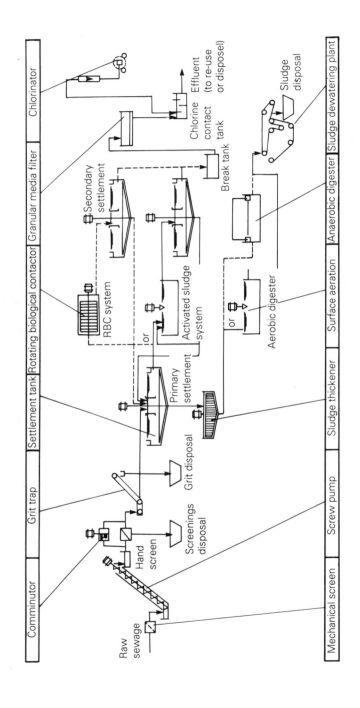

Figure 31.2. Typical large-scale domestic waste-water treatment plant

oxygen from the air. Biological processes convert biodegradable organic materials (BOD) to more microorganisms and carbon dioxide (a gas which escapes to atmosphere during aeration). Most of the microorganisms in the effluent are removed by settlement (secondary settlement) or sometimes by very fine screening. All biological secondary-treatment processes produce more microorganisms which require disposal, sometimes after further treatment (Digestion and dewatering). The excess microorganisms are referred to as secondary sludge (excess, surplus or waste-activated sludge or humus sludge).

Tertiary treatments are the traditional reference for all further processes following secondary treatment. These processes often are those also used for water treatment. The tertiary treatment required, if any, is dependent on the final destination or re-use of the treated effluent, or needed to meet a required analysis (BOD, suspended solids, bacteria etc). This is usually achieved with sand filtration (gravity type or pressure type) sometimes assisted with anthracite. Most are disinfected before disposal and this usually means dosing with chlorine or hypochlorite (sodium or calcium). Sometimes, tertiary treatment is attempted with microscreens (drum screens, fine screening).

A typical secondary effluent of 20mg/l BOD and 30mg/l SS given treatment by sand filtration should produce a filtered effluent of 10mg/l BOD, 10mg/l SS. To improve this quality to 5mg/l BOD, 5mg/l SS further treatment through granular-activated carbon filters would be commonly chosen.

To achieve higher standards than this, chemical dosing and clarification (settlement) using typically polyelectrolytes, alum or ferric chloride, need to be interposed between secondary treatment and sand filtration. A good system could produce 1mg/l BOD, 1mg/l SS.

To produce water of drinking quality with zero concentration of BOD, SS and coliforms (bacteria) it is necessary to interpose ozone treatment between secondary treatment and chemical clarification, and include RO after sand filtration. Activated carbon would not be used as its function is replaced by ozone and RO, and it sometimes generates biological solids which may foul the RO membranes.

Secondary (biological) treatment systems in more detail

Biological treatment is the removal of biodegradable organic materials (expressed as BOD) from water (sewage) and often includes biological oxidation of ammonia to nitrate, sometimes biological reduction of nitrate to nitrogen (a gas which escapes to atmosphere during aeration) and occasionally biological removal of phosphates (arising from detergents) when required. The oxidation of ammonia to nitrate is called nitrification and the reduction of nitrate to nitrogen is called denitrification.

Plate 22. Typical municipal sewage-treatment scheme including trickling biological filters

Organic materials are compounds of carbon and the microorganisms that remove organic materials are called carbonaceous microorganisms. These occur naturally in the environment.

Ammonia (NH_3) and nitrate (NO_3) are compounds of nitrogen (N) and the microorganisms that oxidize ammonia are called nitrifying bacteria. Nitrifying bacteria do not have a significant function in the removal of organics.

If part of the plant is modified to suit, some of the carbonaceous microorganisms can remove nitrate and phosphate. To effect purification as quickly as possible the concentration of microorganisms in the sewage must be increased substantially. For aerobic microorganisms to multiply quickly they need plentiful supplies of food (biodegradable organics) and dissolved oxygen. The organics are already in the sewage and therefore their only requirement is oxygen. Sewage-treatment plant has an aeration facility for this purpose.

Activated-sludge systems

These are a range of biological treatment processes in which the micro-

314

organisms are grown as a free suspension in water which appears in the aeration tank as a rich brown slurry called activated sludge. This process is represented in Figure 31.3. Microorganisms present in the sewage multiply rapidly in the aeration tank due to the mixing and oxygenation (aeration). The mixture of treated sewage plus microorganisms (mixed liquor) flows from the aeration tank to the settlement tank where the microorganisms (activated sludge) settle to the bottom and treated sewage (secondary effluent) overflows a weir. Most of the settled activated sludge is pumped back to the aeration tank to treat more sewage and the extra growth of sludge, which is not required, is removed from the plant (excess sludge).

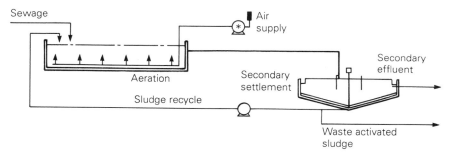

Figure 31.3. Typical activated-sludge process

The conventional activated-sludge process requires primary settlement (before biological treatment) which, in a two-hour retention period, can achieve 30 per cent BOD removal and 60 per cent suspended-solids removal. The remaining dissolved, colloidal and fine suspended organics can be removed in a short aeration period. However the extended aeration modification to the process does not use primary treatment and the period of aeration needs to be considerably extended to be able to accept the whole organic load in the screened sewage.

It would be normal to have average retention periods in aeration for conventional activated sludge and extended aeration of about 6 hours and 24 hours respectively, although these values are not rigid or accurate but depend on the strength of the sewage, the effluent quality required and any consideration of quality and stability of excess sludge.

Contact stabilization is a form of activated-sludge process that attempts to achieve a given effluent quality with smaller aeration tanks. It uses the ability of activated sludge to rapidly absorb organics (BOD) after the sludge has been aerated without food for several hours. It is represented in Figure 31.4. Contact stabilization is considered less reliable than conventional activated sludge or extended aeration in producing a good effluent quality. This is due to factors such as variations in sewage flow creating variations in retention time in the

315

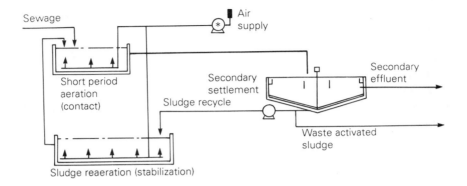

Figure 31.4.   Contact stabilization activated-sludge process

contact tank. Unlike other processes if the retention time is either too long or too short in the contact tank, the effluent quality will deteriorate.

Biological filtration

Biological filters, sometimes called trickling or percolating filters or filter beds, are not filters in the normal sense, that is they do not remove suspended solids by passing the flow through a porous medium. This is a biological contact process of the fixed-film or attached-growth type and, like the activated-sludge process, uses natural aerobic microorganisms to remove biodegradable organic material and, when required, ammonia from sewage. The major differences from the activated-sludge process are:

- the microorganisms grow as a film or slime on the surface of coarse media (traditionally 50mm stones or, more recently, special plastic media);
- the dissolved oxygen is obtained by natural convection of air through the filter; and
- because the microorganisms are fixed, no sludge return is necessary although secondary sedimentation is required to remove the excess microorganisms (humus sludge) that are discharged from the filter.

The depth of the medium is traditionally about 2m although newer designs using effluent recycling can have much greater depth.

Circular filters have rotating sewage distributors driven by the flow of sewage but the distributor can be electrically driven, as is usual for rectangular filters. Coarse suspended solids can choke the filter and therefore primary treatment always precedes the filter.

Rotating biological contactors (RBC)

This process is based on the principle that when an object is rotated about a horizontal axis, partly immersed in sewage, it first becomes wetted and later begins to grow a layer of microorganisms which feed on the organics in the sewage and obtain dissolved oxygen that transfers from the air to the wet film.

The RBC can be a series of discs, a continuous spiral or open plastic matrix. The biological film grows to a thickness dependent on organic load and the speed of rotation. Further growth of film is flushed from the biological contactor into the effluent flow and can be removed by settlement or fine screens.

RBCs require primary treatment (before biological treatment) because they cannot remove suspended solids and the mixing in the RBC chamber is not sufficient to prevent the deposition of primary sludge. The finer suspended solids leaving primary treatment flocculate into larger particles in the RBC chamber and settle with the excess biological solids to form a secondary sludge in the secondary settlement tanks.

The free spaces within the RBC must be large enough to prevent adjacent biological films from meeting and choking the contactor. This is also dependent on the speed of rotation which is usually between 1 and 3rpm.

It is advisable for the RBC tank to be divided into at least 3 stages, sometimes 5 stages, to achieve high-quality effluents. The total contact area required is dependent on the amount and proportion of BOD and ammonia to be removed. The RBC is usually covered to prevent the biological film being stripped off by wind or rain, and to prevent deposition of wind-blown sand.

Large plants have primary and secondary settlement tanks whereas small-package plants usually have pretreatment by septic tank, often including flow equalization, plus a separate secondary settlement chamber or drum filter (microscreen) with return of screened sludge to septic tank.

Small treatment works

For low population application, of the order of one thousand down to individual households, treatment economics dictate a more simplified approach. Figure 31.5 illustrates typical application ranges.

The normal functions of an inlet works are flow measurement, screening and grit removal; the latter two processes are principally to protect downstream pipe-lines and units from blockage. Generally on very small plants flow measurement can be considered as being an unnecessary complication, whilst grit removal should only be contemplated if there is particular reason to suspect high grit concentrations from the sewers.

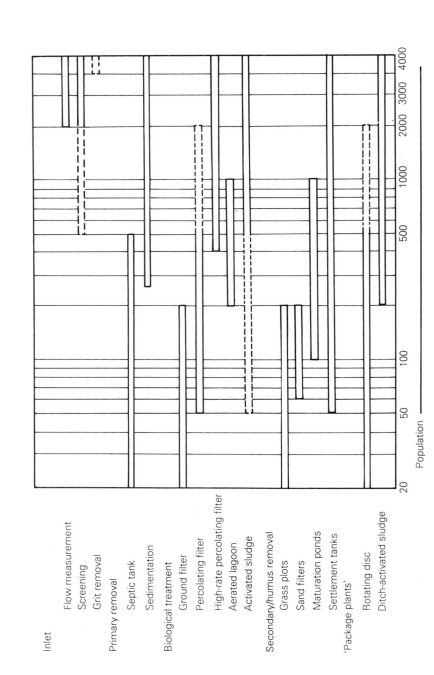

Figure 31.5.   Application of treatment processes vs population

The need for screening is usually more obvious, although the practice should be avoided as far as possible simply because if screens are installed they have to be cleaned. In-line or pump-suction maceration can be a cost-effective alternative.

Septic tanks provide a useful facility for the treatment of sewage on small plants. It should be stressed that cesspools are not treatment units but only storage vessels and now too rarely justified to require consideration.

Multiple-compartment tanks are usually favoured for populations above 30, when about two-thirds of the capacity is provided in the first compartment. For installations serving over about 60 people, there are advantages in having two separate tanks operating in duplicate. Typical arrangements and types of tanks are shown in Figure 31.6.

The arrangement for internal baffles at inlet and outlet are important, since obviously the introduction of crude sewage and the removal of clarified liquid should cause the least possible disturbance of the settled sludge and the surface scum. There are good designs for prefabricated units, either constructed virtually at site from concrete pipes or bought to site ready made in glass-reinforced plastic.

The case for providing septic tanks in two separate parts has been pressed and is perhaps strongest where there is any doubt regarding the regularity or reliability of the tank management and sludge removal. The second tank has significant advantage as a 'catch fence' and should be adopted generally for individual tanks serving 12–100 persons.

The effluent from a small septic tank is most commonly dealt with by discharge to a soakaway system either in the form of a hardcore filled pit or as the size increases by percolation through a field-drain system. The permeability of the soil is most relevant to the system and should be assessed or tested before the system is sized.

Due to their complexity, conventional activated-sludge systems are not normally applied to small treatment works. However, modifications of the system are most effective and can be applied in small plants to unsettled crude

Table 31.3. Design of small treatment works: relative demands of power, land, capital and operator requirement

Process	Land	Capital cost	Power consumption	Operator demands
Activated sludge	1.0	1.00	1.00	1.0
Extended aeration	1.2	1.16	2.20	0.5
Biological filter	4.5	1.83	0.40	0.7
Rotating contactor	1.6	2.22	0.50	0.4
Oxidation ponds	78.0	1.07	0.08	0.1

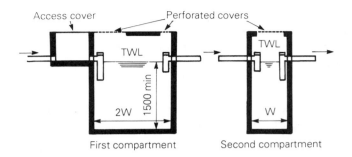

(a) Single tank with provisional second compartment – up to 12 persons

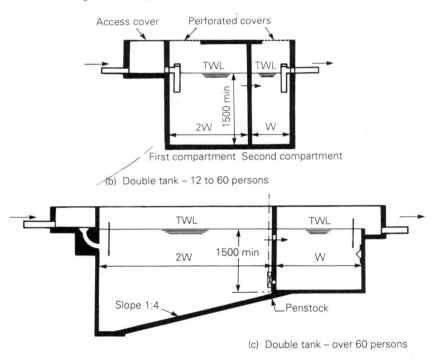

(b) Double tank – 12 to 60 persons

(c) Double tank – over 60 persons

Figure 31.6. Septic tanks

sewage. The common processes are usually contact stabilization or extended aeration systems, and these processes use a variety of aeration devices to mix and aerate the mixed liquor. Both systems have the ability to absorb shock pollutant loads but the simplicity of the extended aeration system has given it prominence.

Packaged rotating biological contactor units are becoming increasingly popular due to the low energy requirement, low profile, low noise level and ease of full enclosure.

32   Treatment of sewage sludge

Introduction

Sludge is a by-product of all sewage-treatment plants. In the raw state, its putrescence results in smell problems when discharged to land. Bacteria and viruses from the sewage are concentrated into the sludge which therefore presents a significant risk to health. Raw sludge is more than 90 per cent water. Dewatering processes can therefore directly reduce transport costs in sludge disposal.

The plant nutrients present in sludge make it a valuable fertilizer if adequately stabilized to reduce smell and infection risk. The organic and moisture contents of sludge make it a good soil conditioner, able to change desert to farmland, forest, grassland or garden.

Sludge can be partly converted to methane gas which is a valuable fuel. Digestion processes reduce the quality of volatile solids, that is organics, to produce a stable residue with reduced smell and levels of pathogenic microorganisms. For more efficient stabilization, for example, of hospital sludges, lime stabilization can be used in addition to either aerobic or anaerobic digestion. Stabilization processes are very popular for conversion of raw-sewage sludges into a form suitable for application for farmland. In the UK for example almost half of the total production of sewage sludge is stabilized, mainly by anaerobic digestion, and used on farmland.

Figure 32.1 illustrates the various routes which can be taken to dispose of sludge.

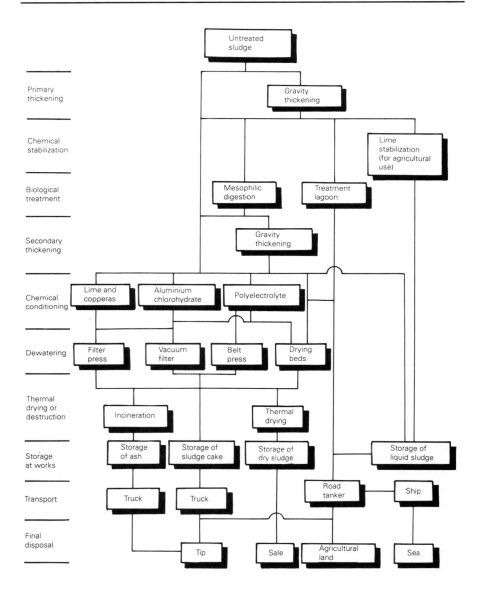

Figure 32.1. Alternative routes for sludge disposal

Anaerobic digestion

Anaerobic digestion is a sludge stabilization process effected by mixing the sludge in a closed tank to isolate the sludge from the atmosphere and thereby produce oxygen-deficient (that is, anaerobic) conditions. This encourages the growth of anaerobic bacteria of which two broad types develop:

- acid-forming bacteria which convert the putrescible material into soluble organic acids and
- methanogenic bacteria which convert the organic acids to methane gas (sludge gas) which bubbles out of the sludge.

In countries where energy costs are high the methane produced by this process is of considerable value for heating and for power generation.

Sludge is typically held in the tank for about 3 to 4 weeks while mixed either by mechanical means, by recirculating pumping, or preferably, by recirculation of the sludge gas. The mixing power is preferably at least 5 watts/m^3 volume. The digested sludge can be thickened by gravity and/or mechanically dewatered to a cake consistency. The final sludge or cake is considered good for use as a fertilizer and soil conditioner.

Anaerobic digestion is dealt with in detail in Chapter 30.

Lime stabilization

Lime stabilization is a simple process. Its principal advantages over other stabilization processes are low capital cost and simplicity of operation. It reduces smell problems and kills disease microorganisms by creating a high pH environment which is hostile to biological activity. Sufficient lime must be added to the sludge to maintain a pH of at least 12.0 for at least two hours. The equipment required includes a contact tank sufficient for two hours mixing time plus two hours settlement time.

The tank must have the facilities to add lime in powder form, efficiently mix the tank contents and, following settlement, to remove the treated sludge to disposal and return top water to the main sewage treatment processes at a slow rate so as not to shock the process with the high pH.

The lime-stabilized sludge readily dewaters with mechanical equipment and is generally suitable for application to agricultural land or disposal at a land-fill site.

For sludges derived from domestic sewage-treatment plants either lime stabilization or a biological digestion process is considered adequate for reduction of disease microorganisms. However, in cases where particularly high efficiency of disease microorganism removal is desirable, for example

323

hospital sludges, lime stabilization can be used following a biological digestion process.

Plate 23.   Sludge-settling system

Sludge conditioning

Sludge conditioning is the pretreatment of sludge to facilitate water removal by a thickening or dewatering process. Conditioning has in the past been achieved in many ways including chemical dosing with lime, iron salts or aluminium salts and by heat treatment. Today almost only one conditioning method is commonly used and that is dosing with polyelectrolytes.

Laboratory pilot-flocculation tests can be performed to determine the correct polyelectrolyte and the optimum dose concentration. This can be checked on-site before the dewatering system is provided with the optimized conditioning system.

Sludge thickening

Thickening increases the solids concentration of the sludge by partial removal of the water content. This can be applied prior to sludge treatment or as an intermediate to reduce sludge volumes and equipment capital cost.

Thickening processes in use include gravity thickeners (sometimes called picket-fence thickeners), dissolved air flotation (DAF) thickeners and centrifuges. By far the most commonly used method is the gravity thickener which on small plants can simply be a hopper-bottomed settlement tank, but on large plants a rotating thickening mechanism is usually employed. All thickening processes either depend on, or work better using, polyelectrolyte dosing. In most applications the thickened sludge volume is less than half, often as little as a quarter of the starting volume.

Sludge dewatering

Dewatering usually removes sufficient water from a liquid sludge to change its physical form to that of a damp solid called sludge cake, although with some secondary sludges in some dewatering processes a thick liquid product is sometimes obtained. The objective of dewatering is to reduce the bulk of sludge for final disposal and reduce transport and other costs.

Almost all mechanical dewatering systems of modern design are dependent on sludge preconditioning with polyelectrolytes. The size of the dewatering plant can often be reduced by using sludge thickening beforehand. There are many dewatering processes available including plate presses, rotary vacuum filters, drying beds, lagoons, centrifuges but the three most popular methods are described below.

Belt press (filterbelt or band press) This machine has been improved several times in the last few decades. The most recent design is particularly effective in producing solid sludge cakes from almost any common preconditioned sewage sludge. It usually gives a sludge cake of around 30 per cent dry solids. The press is a continuous device.

The press consists of two endless belts of woven man-made fibre running over drive rollers and guide rollers like conveyor belts. Pressure-roller systems press the belts together while the sludge is sandwiched between the belts. Water is squeezed out of the sludge which discharges as a continuous sheet of sludge cake when the belts finally separate where a blade is located to assist the sludge cake to separate. The sludge must be well conditioned with polyelectrolyte before entering the press.

Filter press (plate press or plate and frame press) This is the oldest of the dewatering machines which traditionally is manually controlled but was later automated. However manual control is often still preferred. It is thought still to give the driest cake being typically 35–45 per cent solids. The press is a batch device and consists of vertical plates which are held rigid in a frame and which are pressed together between a fixed and moving end. On the face of each individual plate is mounted a filter cloth. The sludge is fed into the press after conditioning with chemicals and passes through feed holes in the trays along the length of the press. The water passes through the cloth, while the solids are retained and form a cake on the surface of the cloth. Sludge feeding is stopped when the chambers are filled. Drainage ports are provided at the bottom of each chamber. The filtrate is collected from these, taken to the end of the press and discharged to drain. The dewatering is complete when the filtrate flow has almost fallen to zero. The feed pumps are stopped, pressure is released and the press is opened one chamber at a time allowing individual cakes to fall out into the collecting vehicle or conveyor.

Drying beds   These usually consist of 50–230mm depth of sand over 200–460mm of gravel with underdrain pipework with open joints or perforations. Sand size is typically 0.3–1.2mm with uniformity coefficient less than 5. Gravel is typically 3–25mm size.

Drying beds are suitable only for digested sludge or sludge from extended aeration-type processes.

The process is of the batch type. Sludge is piped onto the bed over a splash plate and the bed filled to a sludge depth of 200–300mm. Free water drains through the sand and returns to the main treatment plant. The remaining sludge dries by evaporation until, when the sludge cake cracks, it can be lifted manually or on large plants by moving bridge-type lifters. The cake that has just cracked is about 20 per cent solids but continues to dry if left for longer periods.

Land application of sludge

Land application of sludge can have two objectives: the disposal of waste products/residues or the use of nutrients and organic matter to fertilize crops and reclaim soil. In the first case, sanitary landfills or land spreading would be used. In the second case, sludge is considered a resource that can be used as a fertilizer, a soil conditioner or (for liquid sludge) a source of irrigation water.

The dumping of sludge at sea is the subject of much debate and legislation to cover this disposal is being tightened.

33   Effluent problems in ion exchange

Ion exchange is often suggested for effluent treatment (for which it is seldom suitable). The fact that ion exchange itself creates effluent is less widely published. All the same, the running cost of an ion exchange plant comes not only from regeneration chemicals but also from the cost of waste water and of neutralizing waste chemicals. Quite often it turns out that this is the tail which wags the dog when designing plant.

Waste water

The real cost of water is rising, while the real value of money is falling. We should stop saying that someone spends money like water but say instead that some processes waste water like money.

The ion exchange waste water is created during the regeneration process and this waste water has its own cost on top of the chemicals which it carries to the drain. The sewage authority which may accept it for treatment also levies a charge.

The excess acid and caustic used in regenerating a demineralization plant will be greater than the chemical needed by the resin. Regeneration ratios (see Chapter 10) of 2 and 3 for acid and caustic respectively are not uncommon. Since sulphuric acid needs to be diluted to about 2 per cent to avoid calcium sulphate precipitation and caustic soda to about 5 per cent, the water cost is high; additional treated or semi-treated water may be required for regeneration so that the cost of the water escalates.

It is not unusual for an ion exchange unit to use a total of 10BV of water during regeneration. One way to reduce waste water is to use the resin at a higher capacity, so that the 10BV of waste water can be spread over a larger output per cycle. We can do this by increasing the regeneration level, but at higher levels the acid is used less efficiently, so that its proportional cost will rise. As it happens, H_2SO_4 becomes very inefficient at higher regeneration levels so that the resulting gain in water economy is small and is swamped by the increasing cost of acid. On the other hand, a lower regeneration level gives such a low capacity leading to such a high cost for waste water that the gain in efficiency is swamped by the waste-water cost.

For the moment we are ignoring the other effects of changing the regeneration level. Higher capacity reduces the leakage of ions and lowers the capital cost.

Plate 24. An effluent pumping system at a major brewery

Alternative regenerant

To avoid the high waste-water cost, hydrochloric acid (HCl) can be used instead of sulphuric acid (H_2SO_4). Since no problem can arise with calcium sulphate precipitation up to about 10 per cent solution strength, HCl can be used for regeneration and on the mainland of Europe, where it is cheaper than in the UK, it is used frequently. Its cost in the UK is about 3 times that of H_2SO_4 and it requires a more expensive storage and handling system than H_2SO_4 so it is not normally the preferred economic option. There are occasions, however, when a designer has to use HCl because the sewerage authority cannot accept a high sulphate effluent. The small demineralizers in common use in industry nearly always use HCl for cation regeneration because running cost for these small plants is not of prime importance to the end user.

Neutralizing waste chemicals

Regeneration inevitably results in some excess regenerant. It is now a very common requirement that this excess must be neutralized before discharge. The obvious and best way is to let the waste caustic from the anion unit neutralize the waste acid from the cation unit. Here we face the fact that most waters contain bicarbonate. If it is removed as CO_2 by degassing, this leaves the anion unit proportionately less work to do.

Suppose we have to treat a water of 100ppm as $CaCO_3$ total ions and that after degassing, the load on an anion exchanger will be 50ppm as $CaCO_3$ (1meq/l). This consists of the EMA plus the silica and the residual CO_2 left by the degasser.

If these anions are to be removed by a Type II strong base resin, a typical performance would use a regeneration level of 64g/l NaOH (4lb/ft^3), giving a capacity of 510mg/l (1.6lb/ft^3) as $CaCO_3$, and a regeneration ratio of 3.2 results. This capacity would be the same as the cation resin's, so that for every m^3 or ft^3 of cation resin we need only half a m^3 or ft^3 of anion resin in this example.

The anion resin's regeneration ratio of 3.2 is much higher than the cation resin's corresponding ratio of 2.2. We therefore have less anion resin handling less anion load, but at a much lower efficiency.

The cation unit, with a load of, say, 100ppm produces the equivalent of 120ppm of waste acid. The anion unit with its load of 50ppm produces 110ppm of waste caustic (all as $CaCO_3$) so that the two effluents nearly, but not quite, neutralize one another.

Service water requirement

There is another complication, which is that the water used to regenerate the anion unit must be free of hardness, or the high pH conditions would cause precipitation. Rather than install a separate softener to supply regeneration water, we use semi-treated water from the cation unit. The design of the cation unit must therefore be scaled up to meet the additional demand, and this in turn creates more waste acid. (The same service water problem arises at every step of a treatment train. Coagulation tanks have to provide for filter backwash water, filters for cation waste, cations for anion waste and so on. At each stage the losses multiply one another. Overall, a treatment train can easily waste $350m^3$ for every $1000m^3$ net output.)

For example, an anion unit needs about 10BV of semi-treated water for its regeneration and rinse. Its capacity is 510meq/l which, when treating a water of 50ppm anions as $CaCO_3$ (1meq/l), gives a run of 510BV. The anion service water is therefore 10BV in 510BV, or 2 per cent, so everything on the cation unit has to be scaled up by a factor of 1.02. Then its waste acid will in fact be, not 120ppm, but 122.5ppm equivalent.

This hardly seems worth bothering with, but then our sums are based on a very thin water. The percentage of service water depends on the ionic contents of the water; on a thick London water analysis the anion service water would be 8 per cent.

Balancing the cation/anion pair

In the example above, the units were not quite self-neutralizing. By dint of some lengthy arithmetic it is possible to adjust the regeneration levels to make it so. There are two advantages to a self-neutralizing plant: it is clearly more efficient to use all the chemicals as regenerants first and so get some benefit from them, rather than dosing neutralizing chemical straight into the effluent; and a self-neutralizing plant will stay that way unless the raw-water composition varies. This can make it possible to avoid separate effluent neutralization gear with dosing pumps and pH meters, which are expensive and troublesome.

It is important to remember, however, that neutral in this context means that the bulked effluent from each regeneration will contain very little excess acid or alkali. It does not necessarily mean that the effluent is always within a narrow pH band.

330

34 'Neutral effluent' — how neutral is it?

We now go on to discuss the problems which arise out of having to meet a pH limitation and will ask whether such pH limitations have much useful meaning.

A typical plant

In order to get a feel for the problem, let us work through an actual design example. Let us assume that we are to design a cation/anion pair with a neutral effluent, to yield 150m^3 (33 000g) of demineralized water between regenerations. A model solution giving the capacity and regeneration conditions to satisfy this requirement is set out in Table 34.1 which shows that we end up with a total of 35m^3 (7700g) of bulked effluent, 23.3 per cent of net output, at the end of each cycle. On paper, the acid and alkali neutralize one another correctly to within 0.1 per cent.

Accuracy of regenerant measurement

It is one thing to calculate chemical doses correct to four significant figures (so easy with an electronic calculator on your desk!); it is quite another to make a plant actually operate with such accuracy. The makers of chemical metering pumps claim accuracies between ± 2–5 per cent, depending on the quality of the pump. As pumps wear with age, this accuracy deteriorates. Measuring tanks are more accurate; in the sort of size plant we are now considering, a

Table 34.1. Typical demineralization plant chemical design

	Cation unit	Anion unit
(1) Output per cycle (cation unit output includes anion service water)	$160.5m^3$	$150m^3$
(2) Ionic load per cycle = (1) × ions in analysis	1310g equiv	612g equiv
(3) Regeneration ratio	1.98	3.10
(4) Capacity of resin	480meq/l	580meq/l
(5) Resin volume required $= \dfrac{(2)}{(4)}$	$2.73m^3$	$1.06m^3$
(6) Regenerant applied = (2) × (3)	2594g equiv (acid)	1897g equiv (caustic)
(7) Excess regenerant = (6) − (2)	1284g equiv	1285g equiv
(8) Bulk of effluent produced*	9 Bed Vols $= 24\frac{1}{2}m^3$	10½ Bed Vols $= 10\frac{1}{2}m^3$

*The number of Bed Volumes of effluent which must be bulked for neutralization depends on the plant design and the care with which the plant is operated. These and other conditions dictate whether the backwash and sometimes some of the rinse water are to be re-used, whether they can be put down the drain without neutralization or whether they have to be bulked with the effluent. The figures in this example are an arbitrary choice, merely to give us something as a basis for working out the example.

batch measuring tank can work with a reproducibility of ± 1 per cent, but of course this has to be set by trial and error at the commissioning of the plant.

The metering of regeneration chemicals is therefore going to involve significant deviations from the calculated figures. The greatest upset to a neutral effluent would result from the errors going in opposite directions, that is if the acid dose is high while the caustic dose is low, or vice versa. In our example the total quantity of acid plus caustic is 2594g equiv of acid + 1897g equiv of caustic, totalling 4491g equiv of regenerant (equivalent to 224.6kg as $CaCO_3$). A 1 per cent error in metering the chemicals can therefore produce an imbalance of up to 2.2kg as $CaCO_3$ of either acid or alkali, depending which way the errors went, and a 5 per cent error would produce an imbalance of up to 11.2kg as $CaCO_3$.

When bulked, our effluent has a total volume of $35m^3$. In this volume, 2.2kg as $CaCO_3$ gives a concentration of 63ppm excess acid or alkali, and 11.2kg as $CaCO_3$ gives 320ppm. In the absence of buffering, these concentrations would give quite strongly acid or alkaline solutions.

Accuracy of the water analysis

The raw-water analysis is the rock on which a plant design stands. Quite often this turns out to be very shifting sand. The standard methods of water analysis are capable of giving results in which the major constituents are all correct to within 1ppm. Analytical chemists are very touchy if you call into question the accuracy of their results, but a series of investigations by the Water Research Association has officially confirmed what we always knew, that errors as high as 20ppm in a single constituent can be reported by laboratories which are supposed to be experienced in water analysis. Errors of 10ppm are quite common.

Suppose in our example there is one such error of 10ppm; suppose that the real anion concentration in the raw water is not 204ppm, as the analysis says, but that it is actually only 194ppm (3.88meq/l). Then the real anion load per cycle will be less than we have calculated on paper, and therefore the spent anion regenerant will contain correspondingly more unused caustic soda. In our example this error of 10ppm anions will yield a surplus of caustic soda equivalent to 1.5kg $CaCO_3$.

We could insist on more reliable water analyses if there were much point in having them. The fact is, of course, that natural waters vary from time to time, so that from the practical point of view the analytical error probably falls within the natural variation of the real analysis (which does nothing to excuse the shocking errors which the survey showed). River waters can change quite sharply between summer and winter, and even well waters tend to vary seasonally. (As an extreme example, a demineralization plant in Rugby, England, failed in the winter after a nearby motorway was opened. It turned out that salting the motorway in icy weather trebled the chloride in the shallow well which supplied the plant.)

Public supplies are often blended from different sources; when the mix changes, the analysis changes with it. If the proportions of ions stay the same, then a change in analysis will not affect the neutralization of effluent, though the run length will of course change. But if the constituents change in proportion to one another, then excess of acid or alkali will appear in the bulked effluent.

Buffering: the effect of bicarbonate

Just for once there is a slight suspension of Murphy's Law, and a helpful natural factor in this problem. Both raw water and the spent regeneration contain buffering. The regeneration and rinse of the cation unit are done with raw water, which becomes strongly acidic in the process; the raw water alkalinity is therefore all converted into CO_2. Some of this CO_2 may be lost to the atmosphere, expecially if the effluent splashes about in open drains, but at least some will stay in solution.

On the anion side the buffering comes from CO_2 accumulated on the resin during the run. The anion unit is fed from the degasser sump with a water which, we said, contained 10ppm of CO_2. This is taken out with the mineral anions and reappears as Na_2CO_3 in the spent regenerant. (The regeneration and rinse of the anion unit are done with degassed water, and the amount of buffering left in that is so small that it can be ignored.) Both the cation and anion effluents therefore contain some buffering chemical. Reverting to our example, we can calculate how much of it there might be.

Our raw water contains 229ppm bicarbonate, as $CaCO_3$. When acidified, this converts to 458ppm CO_2 as $CaCO_3$. As a pure guess we might assume that 200ppm of this actually remains in the $24.5m^3$ of cation effluent reaching the effluent sump, making a quantity of 4.9kg CO_2 as $CaCO_3$.

We assumed that there would be 10ppm of CO_2 left in the water by the degasser, which would be removed by the anion resin. A cycle of $150m^3$ therefore puts 1.5kg as $CaCO_3$ on to the resin. When this is regenerated off with caustic soda, it reappears as 1.5kg as $CaCO_3$ of Na_2CO_3.

Let us assume that (more by luck than by judgement) we have actually hit off perfect balance between acid and alkaline effluents, so that the strong cations and anions neutralize one another exactly. Then the CO_3^{--} from the Na_2CO_3 will also react with the CO_2 (see Chapter 12) to form bicarbonate. We have 1.5kg as $CaCO_3$ of CO_3^{--}, which takes 1.5kg of CO_2 from the 4.9kg available, to produce 3kg of HCO_3^- and leave 3.4kg of CO_2 behind (all these are as $CaCO_3$). The resulting ratio of alkalinity/free CO_2 is then 3.0/3.4 which equals 0.88. This gives a pH of 6.6.

In practice we will not have exact neutralization of the strong cations and anions, but if we have extra acid it will first convert HCO_3^- to CO_2, and if we have extra caustic it will do the reverse. If the extra acid or alkali quantity is less than the HCO_3^- or CO_2 available, then the pH of the effluent will stay within the range of pH4–8 and will obey the graph in Chapter 12.

Table 34.2 compares the actual quantities which are likely in our worked example, so that we can see how useful this effect might be. Now that we see all these numbers side by side, it is clear that we must expect excesses of chemical in the bulked effluent which are greater than the available buffering chemicals. In this example, even with the greatest care, we cannot guarantee an effluent between pH4-8 unless we monitor and correct the pH after bulking the effluents.

London water, which forms the basis of this example, has a TDS which is above average so we might look at the corresponding problem as it would be

Table 34.2. Typical components in effluent which affect the pH

HCO_3^- in bulked effluent, assuming exact neutralization of regenerants	3.0kg as $CaCO_3$
CO_2 in bulked effluent, assuming exact neutralization of regenerants	3.4kg as $CaCO_3$
Excess of acid or alkali due to 1% error in chemical measuring gear	2.2kg as $CaCO_3$
Excess of acid or alkali due to 5% error in chemical measuring gear	11.2kg as $CaCO_3$
Excess of acid or alkali due to 10ppm error in raw water analysis	1.5kg as $CaCO_3$

on a thinner raw water. As an alternative example, let us take a water containing 50ppm of anions. If we assume that we put it through a unit of the same size as our first worked example with 204ppm of anions, then we shall get a run which is about four times as long, say $600m^3$ between regenerations. But the CO_2 after the degasser is going to be the same in both cases, which is 10ppm, so that this time the amount of CO_2 going on to the anion resin is 6.0kg per cycle.

On this other analysis, assume that the raw water contains 50ppm of alkalinity, so that the cation service-water will only contain 100ppm CO_2. Even if we took great care to keep all of this in solution, then our $24.5m^3$ of cation effluent will only contain 2.45kg of CO_2, which is not enough to react with the 6.0kg of CO_3^{--} coming off the anion resin with the caustic regeneration. The bulked effluent would contain surplus CO_3^{--}, its pH would be above 8.0 and the surplus carbonate would precipitate out as calcium carbonate.

In this case we have to use a little more acid or a little less caustic in order to balance the final effluent. An extra 5kg of acid as $CaCO_3$, for example, would go through to the bulked effluent and would then convert 5kg as $CaCO_3$ of CO_3^{--} ions to 5kg of HCO_3^-. This leaves 1kg of CO_3^{--} to react with 1kg of CO_2 to give another 2kg of HCO_3^-, making 7kg HCO_3^- in all, and leaving 1.45kg of free CO_2 in the bulked solution. Then the alkalinity/free CO_2 ratio is 7.0/1.45, which equals 4.8 and corresponds to a pH of 7.5. The bulked effluent now contains about twice as much buffering as it did in our first example, thanks entirely to the extra CO_2 picked up on the anion resin.

This suggests a way of dealing with the situation shown in our first example where the buffering was too small to take up the expected errors. Suppose we had deliberately built a degasser to operate at reduced efficiency, so that it left a residual of 40ppm CO_2 in the degassed water. In a $150m^3$ cycle this would put 6kg of CO_2 in the anion resin. Suppose, in addition, that we took great care to avoid loss of CO_2 from the cation effluent, so that all its 458ppm remained in solution as it entered the effluent tank, then in $24.5m^3$ of cation effluent we would have 11.2kg of CO_2. Without going through the arithmetic yet again, when these two neutralize one another we obtain 12kg HCO_3^- and 5.2kg CO_2, at a pH of about 7. Now the buffering should be enough to keep a well-designed and well-operated plant within the pH range of pH4–8, without the nuisance and cost of monitoring and trimming the bulked effluent pH.

There is of course a price to pay for this, which is the extra caustic soda needed to regenerate the extra 30ppm of CO_2 off the anion resin. On a small plant this might be a small price to pay for the convenience of a self-neutralizing effluent without monitoring and trimming.

In practice this device is not used today. Ironically, one reason why this is so is that it is not easy to design a degasser with a fixed and deliberate degree of inefficiency. However, the art of designing plant for neutral effluent is still quite young, and we cannot say that we have fully explored all the possibilities.

Effect of dilution

Dilution must bring pH nearer to neutral, and all effluents are diluted somewhere along the line. If the diluting water contains bicarbonate, then the neutralizing effect can be very useful. For example, the London water taken for our worked example is quite a massive source of buffering, as it contains 229ppm bicarbonate and about 40ppm free CO_2.

If a works happens to have a waste stream of water available for blending with the ion exchange effluent, this could (with luck) guarantee a total effluent within the desired pH range.

What is 'neutral'?

From our worked examples we can see the issues involved in practical design, which can be summarized as follows:

- It is relatively easy to design a demineralization plant whose cation and anion effluents will roughly neutralize one another when bulked.
- Inevitable deviations from the design figures arise from errors in chemical dosing and variations of the water analysis.
- CO_2 and bicarbonate buffering reduce the pH swings which result, but the amount of buffering is not normally enough to cover them.
- To keep pH within specified limits therefore needs equipment to monitor and trim the bulked effluent. This is costly to buy and troublesome to operate and the cost and nuisance become relatively greater on smaller plant.
- Dilution of the effluent helps to bring the pH nearer to neutral, especially if the dilution water has a high alkalinity.

Against this background it is worth asking why we have to neutralize effluent at all. Very strong acids and alkalis would of course react with the contents of drains and sewers, but the near-neutral solutions which we are now considering are harmless from that point of view. The real problem is one of corrosion.

The relevance of pH

The actual pH range which sewerage and river authorities will accept varies from place to place, and actual values can appear arbitrary. The fact is that pH is not really a very good parameter for the purpose.

A heavily buffered effluent just above pH4, for example, can in reality do more damage than an unbuffered effluent at pH3.5. For example, let us compare an effluent at pH4.5, buffered with 1000ppm CO_2 and 7ppm HCO_3^-, with an unbuffered one at pH3.5. Diluted with an equal volume of London water, the buffered effluent gives us 520ppm CO_2 and 118ppm HCO_3^- at pH6, which is still quite corrosive. The unbuffered effluent at pH3.5, on the other hand, only contains 20ppm of free mineral acid and when this is diluted with an equal volume of London water the resulting blend contains 40ppm CO_2 and 104ppm HCO_3^-, at pH7.1. In the long term, therefore, the unbuffered effluent is potentially less corrosive even though at the point of discharge it is the more strongly acid.

This sort of argument makes it impossible to define corrosiveness. pH at least has the virtue of being easily defined and easily measured (or it would have, if pH meters were more reliable. To keep them working well needs conscientious maintenance).

What is 'non-neutral'?

Now that we have shown the weakness of pH as a criterion, it is worth considering what would happen if we discharge a non-neutral effluent, that is, if we designed self-neutralizing plant to give a roughly neutral effluent but to omit the final pH monitoring and trim.

Table 34.2 summarized the results of our first worked example. It showed that the worst condition would give us an excess of nearly 13kg as $CaCO_3$ of acid or alkali, of which 3kg or so would be taken up by bicarbonate buffering. This leaves us with an excess of about 10kg of free acid or alkali in a bulk of 34m^3 of effluent.

To dump this down a sewer sounds quite alarming, but then this bulk only arises at each regeneration cycle, which means in practice at intervals of 8 to 24 hours. At the worst, therefore, we could arrange to let this down slowly over 7 hours (leaving 1 hour as a safety margin). We should then release

5m^3/h containing at worst 1.4kg as $CaCO_3$ of acid or alkali (1100g/h containing 3lb). It takes quite a small flow of dilution water to render such an effluent harmless, either on the works or perhaps in the sewer itself. For safety, we could change the regenerant doses to give a deliberate excess of caustic, so that instead of having 1.4kg/h of acid or alkali, the swing would be between 0–2.8kg/h of alkali, which reduces the danger of corrosion.

All the examples given are calculated on the basis of an ion exchange plant with a cation unit containing 2.73m^3 of resin, which is something like a 2000mm-diameter unit, quite a medium-sized plant. Both the volume and the chemical contents of the ion exchange effluent after each regeneration are roughly in proportion to the volume of resin in the unit, so that a 1200mm-diameter unit, for example, would produce less than half the effluent. On the other hand the cost and nuisance of equipment to monitor and trim the pH of an effluent varies very little with the scale of the operation. It is therefore on the smaller plant where we would have the greatest benefit from being able to discharge a non-neutral effluent. Of course this still means designing the plant to be self-neutralizing. It also means providing the effluent bulking tank for the two streams to mix. (There are ways of using additional ion exchange units to do some of this work and so reducing the size of the tank.)

Summary

A demineralizing plant can be made self-neutralizing but the effluent is bound to vary somewhat unless it is monitored and pH trimmed by dosing. However, a pH criterion is not necessarily the best measure of what the sewerage authority really wants; there are cases when it might be better to discharge a roughly neutralized effluent in a carefully regulated trickle rather than dumping a pH-corrected effluent at an uncontrolled rate. (However, this argument is weakened because of the great difficulty experienced in monitoring effluent flow.)

These points are worth discussing with the sewerage authority. Industrialists often do not know just why they are being asked to control the pH of their effluent (and as a matter of fact some sewerage authorities aren't all that clear about it either). In discussion it might turn out that the rate of discharge, or the total amount of surplus chemical per day, are the important elements, and this might simplify the design and operation of the ion exchange plant.

Appendix 1 Units of measurement in water analyses

The confusion about units of measurement was quite bad enough before metrication days. Things have not got easier since. It is therefore worth taking a little time to look at this subject critically.

Basically there are three kinds of units in common use:

- those which actually measure the concentration in which a substance is found;
- those which measure its chemical equivalence; and
- those which measure some fairly arbitrary property, sometimes on a wholly arbitrary scale.

Units of concentration

The basic and commonest unit of concentration is the ppm (part per million). It means that there is one unit weight of the substance to a million unit weights of the solution, and it is therefore a straightforward ratio, without units of its own, and (happily) unaffected by metrication.

A commonly used metric equivalent is the mg/l (milligram/litre) which in water and dilute solutions is in practice interchangeable with the ppm. In theory they are not exactly the same, the mg/l being a measure of weight per volume, but for ordinary waters this need worry no one. The mg/l is identical to the g/m^3 (gram/cubic metre), and this latter unit is the official SI unit, though never used in practice.

Very high-purity waters are described in ppb (parts per billion), where the b is the US billion, that is 10^9. The metric equivalent is mg/l (microgram per litre), a unit which avoids the possible confusion as to whether it is the British or the American billion which is being used. The simple description not open to error is 'parts per 10^9'. As we get into ultra-pure water the ppt or 'part per trillion' begins to appear. Its pedantically correct equivalents are 'ng/l' or 'parts per 10^{12}'.

All these units measure the actual concentration of a substance 'as such', to avoid confusion with units of equivalent concentration, some of which are very similar.

Units of equivalent concentration

Actual concentrations of substances are not often of direct use to us; what we want to know is how reactive the water is. For practical use we frequently want units which give the analysis on some common basis, that is, units in which the constituents are shown in the form of some equivalent concentration.

The ions in water react by virtue of their electrical charges. Any measure of equivalent concentration therefore has to be based on the weight of each ion and the number of charges on it. The weights of ions being fantastically small, the best way of measuring them is as a ratio.

Atomic weights

The unit for weighing ions (the same unit is of course also used for atomic and molecular weights) is the H^+ ion, which is the lightest of them all. The Na^+ ion, for instance, weighs 23 times as much, and its ionic weight is therefore 23. The Ca^{++} ion weighs 40 but, as it has two charges, its equivalent weight per charge is 20. The SO_4^{--} ion contains an atom of S weighing 32, and four of O weighing 16 each; the total weight of the ion is therefore 96, and the equivalent weight (per charge) is 48.

Ions always combine in quantities proportional to these equivalent weights. For instance, $CaSO_4$ is made up of a Ca^{++} ion and an SO_4^{--} ion and takes these two ions in the proportion 40:96 (which is the same as 20:48). On the other hand when Na^+ ions combine with SO_4^{--}, because the Na^+ only has one charge, it takes two ions of Na^+ per SO_4^{--} to make Na_2SO_4, and the proportion by weight is (2×23): 96, which is the same as 23:48.

If we know the concentration of an ion measured in 'ppm as such' and divide this by the equivalent weight of the ion, then we get a measure of concentration in terms of its power to react with other ions. This is called the meq/l, or mg equivalent per litre. A solution of an ion at 1meq/l will just react completely with the same volume of a solution of another ion at 1meq/l. Also, the total cations in a solution must equal the total anions, provided each ion is measured as meq/l.

Using the meq/l unit

The meq/l could also be called the 'equivalent per million', or epm, which is a ratio and therefore dimensionless. The following example shows how useful this kind of unit is.

Suppose we make up a solution of 200ppm NaCl and 100ppm $CaSO_4$. The analysis of the resulting solution can be written in several ways:

1. We can write down the concentration of each constituent salt in ppm:

$$
\begin{array}{ll}
\text{NaCl} & 200\text{ppm} \\
CaSO_4 & 100\text{ppm}
\end{array}
$$

This tells us how the solution was made up, but is not much use in telling us about its behaviour as a solution.

2. We can divide the salts into their constituent ions by dividing the concentration of each salt by its equivalent weight, and then multiplying by the equivalent weight of the ion, to get the concentration in 'ppm as such':

$$
\begin{array}{ll}
Na^+ & 78\text{ppm as such} \\
Ca^{++} & 29\text{ppm as such} \\
Cl^- & 122\text{ppm as such} \\
SO_4^{--} & 71\text{ppm as such} \\
\text{Total} & 300\text{ppm}
\end{array}
$$

This tells us the weight of the individual ions, but in practice these are still no use at all, except that the total gives us the total dissolved solids (TDS), which gives a general indication of what kind of water we are dealing with.

3. We can take the equivalent concentration of the constituent ions in epm, simply by dividing the concentration of each salt by its equivalent weight

$$
\begin{array}{llll}
Na^+ & 3.4\text{meq/l} & Cl^- & 3.4\text{meq/l} \\
Ca^{++} & 1.5\text{meq/l} & SO_4^{--} & 1.5\text{meq/l} \\
\text{Total} & 4.9\text{meq/l} & \text{Total} & 4.9\text{meq/l}
\end{array}
$$

This tells how much of each ion is present in terms of its ability to form chemical combinations. It shows (what, in this case, we knew from the way the solution was made) that the Na^+ and the Cl^- are equivalent to one another, as are the Ca^{++} and the SO_4^{--}. It also shows that Na^+ and Cl^- account for 70 per cent of the salts in the solution, when measured by their ability to react, and finally it shows the total ionic strength of the solution, and that the cations and anions are in total equal to one another, which they must be if the analysis is correct.

The main drawback to the meq/l unit is that in a water analysis of normal accuracy it runs to one or more decimal places, which makes it a little unwieldy.

Similar units

The meq/l has other units which are identical to it, just like the ppm. These are: the g equiv/m^3 (the gram equivalent/cubic metre), which is the SI unit, and the mval/l (millivalency/litre).

Because of the convenience of units based on equivalent concentration, analytical chemists have a very similar system, based on normal (N) solutions.

N solution contains a gram equivalent per litre, which is the same as 1000meq/l. (The N system, like the meq/l is a weight/volume system which is important at the concentrations used in other branches of chemistry. At N concentration, the density of solutions is significantly greater than 1.000.) N/50 solution, which is commonly used in water analysis, is the same as 20meq/l.

ppm as CaCO$_3$

An older way to measure equivalent concentration is to use the 'ppm as CaCO$_3$', in which every ion is calculated as the chemically equivalent concentration as if it were CaCO$_3$. The use of CaCO$_3$ comes from the fact that its molecular weight is exactly 100, but as it is supposed to dissociate into Ca^{++} and CO$_3^{--}$, its equivalent weight is 50. Continuing with our previous example:

4.   Our same water analysis is now calculated by taking each of the constituent salts in ppm as such, dividing by its equivalent weight, and multiplying by 50:

Na$^+$	171ppm as CaCO$_3$	Cl$^-$	171ppm as CaCO$_3$
Ca^{++}	74ppm as CaCO$_3$	SO$_4^{--}$	74ppm as CaCO$_3$
Total cations	245ppm as CaCO$_3$	Total anions	245ppm as CaCO$_3$

The information this gives us is exactly the same as that which is given by the meq/l unit; as a matter of fact this is merely the meq/l analysis multiplied by 50. The advantage of this unit is that the concentration of each ion, and that of the total ions, is of the same order of magnitude as the concentrations 'as such', and this gives a more direct idea of what this water is really like. For instance, we know that the TDS is 300 (see no.2 p.341); in 'ppm as CaCO$_3$' the total ions are 245ppm as CaCO$_3$. The other advantage is that in normal water analyses we are dealing in whole numbers rather than decimals.

On the other hand, the 'ppm as $CaCO_3$' is easily confused with 'ppm as such', and many slapdash analysts fail to say which system they are using even when they are using both systems within one analysis, as often happens. The other problem is that, while $CaCO_3$ is splendidly handy in having an equivalent weight of exactly 50, it is not actually soluble in that form and the whole concept is likely to cause confusion when we come to measure bicarbonate or carbonate as 'calcium carbonate'.

The way to avoid confusion is to remember that the 'as calcium carbonate' bit is just a convention, and in practice is simply a 50-fold multiplier.

Kilograin per cubic foot

One hardy survivor from a past age, rarely used these days, closely allied to the 'ppm as $CaCO_3$' is the $kgrn/ft^3$ (kilo*grain* per cubic foot; it is very important to get this spelling right to avoid confusion with the kilogram (kg)). It measures the chemical capacity of an ion exchange resin, and 'as $CaCO_3$' is always implied even when not actually stated. It survives especially in the USA where some purchasers still work out their resin needs by the cubic foot, and because the usual commercial resins have capacities between 10 and $40 kgrns/ft^3$, so that it yields handy numbers. The meq/l will serve equally well for resin capacity and, if it is used for water analysis as well, the calculation of ion exchange resin performance becomes very simple indeed.

Table A.1 summarizes all these units.

Table A.1. Units of measurement: conversion table

ppm* $CaCO_3$	meq/l†	$kgrn/ft^3$ $CaCO_3$	lb $CaCO_3$/ ft^3
1	0.02	0.436×10^{-3}	62.3×10^{-6}
50.0	1	21.8×10^{-3}	3.11×10^{-3}
2290	45.8	1	0.143
16 100	321	7.0	1

Notes: * or mg/l $CaCO_3$
$\dagger$ or epm, or mval/l
NB 1 meq/l $\equiv$ N/1000
The column showing lb $CaCO_3/ft^3$ is added to assist in calculation of regenerant doses.

Arbitrary units

Turbidity, colour, smell, organic content, these are properties of water which are difficult to measure and to relate to some basic physical or chemical quantity. On the other hand pH and conductivity are easy to measure but do not tell us directly what the water actually contains.

The tests which are used to determine turbidity and colour are described in British Standard 2690, Part 9; they compare these qualities with those of standard synthetic solutions but, as there are different kinds of colour and different kinds of turbidity in natural waters, the results cannot be of much use except as a general guide. The estimation of smell defies analysts, who are reduced to using their noses.

Organic matter is most commonly measured by its power to take up oxygen from oxidizing chemicals or from the air by various methods. These tests are not easy to carry out with precision; they tell us nothing direct about the type and little about the quantity of organic matter present. The simplest and therefore commonest test used in industry is the amount of potassium permanganate ($KMnO_4$) taken up (see also BOD and COD) but without some additional knowledge of the type of organic matter which the water is likely to contain this is virtually useless. As even the $KMnO_4$ test is not all that easy to perform accurately, this may be just as well.

BOD and COD

There are two measurements of the oxygen demand of polluted waters. Biological oxygen demand (BOD) is a measurement of the oxygen starvation of organisms in the water. Chapter 28 gives an explanation of how this is measured and reported.

Chemical oxygen demand (COD) is a measurement of the oxygen demand in a solution from both organic and inorganic constituents. Both measurements are required when designing effluent treatment plant unless the waste stream is known to be non-biological.

pH and conductivity

pH is a simple and exact test to measure the H^+ concentration in the water, and hence the acidity or alkalinity of the solution. However, the result must be read together with the CO_2 and bicarbonate concentration to give a true picture.

Conductivity is even simpler and more exact than pH but, as different ions have different powers of conveying currents, a conductivity test will give an exact knowledge of a water only if one pair of known ions is present or if all

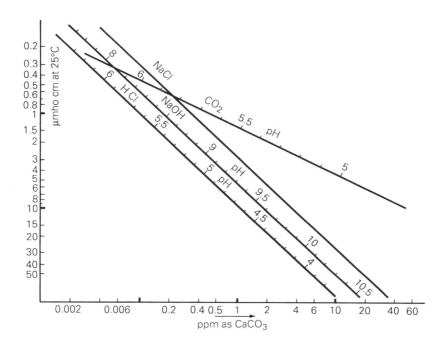

Figure A.1. Conductivity of ions vs concentration

ionic concentrations except one can be measured by some other means. Figure A.1 gives an idea of the conductivities of solutions, which are additive. For example, suppose a water which contains 0.5ppm NaCl plus 0.5ppm HCl (both as $CaCO_3$), whose conductivities are $1.6\mu S$ and $5.1\mu S$ respectively; the conductivity of the water will therefore be $6.7\mu S$.

Most ions have conductivities of similar magnitude except H^+ and OH^- ions whose concentrations can be measured by pH. This means that most salts, acids and alkalis have conductivities roughly similar to those of NaCl, HCl and NaOH as shown in Figure A.1. CO_2, however, is rather different and is shown separately. If both the pH and conductivity are measured, therefore, it is possible to get quite a rough estimate of the total ions present. This is especially useful in examining nearly pure waters such as those produced by demineralization plants, provided we can assume they contain no CO_2. The examples which follow show how this is done.

Example 1 A water coming out a cation/anion demineralization plant has a conductivity of $4.5\mu S$ and a pH of 9.2.

On the NaOH line, the pH of 9.2 corresponds to 0.7ppm NaOH as $CaCO_3$ and from the same line this concentration gives rise to a conductivity of 3.5μS. There is therefore a conductivity of 1μS due to salts, and on the NaCl line this gives 0.4ppm NaCl as $CaCO_3$. This water therefore contains:

$$Na^+ \quad 1.1\text{ppm as } CaCO_3$$
$$Cl^- \quad 0.4\text{ppm as } CaCO_3$$
$$OH^- \quad 0.7\text{ppm as } CaCO_3$$

Example 2 A water coming out of a mixed bed has a conductivity of 1.0μS and a pH of 5.7.

On the HCl line, pH5.7 corresponds to 0.1ppm HCl as $CaCO_3$ whose conductivity is 0.85μS. There is therefore a conductivity of 0.15μS due to neutral salts and the NaCl line shows this is corresponding to about 0.06ppm.

However, if there is a chance that the sample has been contaminated by CO_2, then the acidity may be entirely or in part due to CO_2. This raises a conductivity which is very similar to that of HCl, so that the conclusion about the content of neutral salt is not seriously affected.

Organically contaminated waters contain organic acids which tend to pass through ion exchange plants, and these could also cause a pH and conductivity similar to that of CO_2.

The amount of CO_2 produced by bicarbonate

A lot of confusion can arise when ionic concentrations and CO_2 are measured as 'ppm as $CaCO_3$'.

The point to hang on to is that the ions and molecules of CO_2, $CaCO_3$, HCO_3^- and CO_3^{--} each contain one atom of carbon. When they change into one another, any one of them can only become one of the others, but starting with CO_2 there is the possibility that it will go to monovalent HCO_3^- or divalent CO_3^{--}. The convention of measuring 'as $CaCO_3$' assumes that it will go to divalent CO_3^{--}, though in HCO_3^- it has only gone half way. Hence the rule that *1ppm as $CaCO_3$ of bicarbonate alkalinity produces 2ppm as $CaCO_3$ of CO_2*. Chemists who work in N units or mols do not have this problem.

Appendix 2   Glossary

*Words in **bold type** are defined elsewhere in this glossary.*

ABS stands for two things: in chemical pipework etc, it is a material of construction called Acrylonitrile Butadiene Styrene. In water analyses it stands for Alkyl Benzene Sulphonate, which is the commonest class of detergents.

ACIDS are substances which **dissociate** and release H^+ **ions** into the solution.

ACRYLIC RESINS are anionic or **weakly acidic** ion exchangers: they are not based on **polystyrene.**

ACTIVATED CARBON is made from coal or coconut shells, etc, to acquire a highly porous structure which adsorbs organic molecules, especially smaller ones. It also removes free chlorine from water.

ACTIVE GROUPS or **SITES** are really fixed ions bolted on to the **matrix** of an **ion exchanger.** Each active group must always have associated with it a **counter-ion** of opposite charge.

ADSORBENT is a material whose surface attracts substances and causes them to settle and adhere to the surface. **Activated carbon** is one common example.

AFFINITY is the keenness with which an **ion exchanger** takes up and holds on to a **counter-ion.** Affinities are very much affected by the concentration of the

electrolyte surrounding the ion exchanger. See also **selectivity**.

AIR MIX Bubbling large quantities of air through an ion exchange column causes the resin to turn over and mix intimately. This is done in **mixed beds** and **Unimix.**

AIR SCOUR in sand filters improves **backwash** as it produces more shear and agitation of the bed than is possible by backwash. Alternatively this duty can be performed by **surface washing,** using pressure jets in or on the surface of the bed, or by mechanical rakes which break up the accumulated dirt on the surface of the filter bed.

ALKALINITY in natural waters is the amount of **bicarbonate** present. Actually the word alkalinity also includes hydroxyl and carbonate, but these are extremely rarely found in natural waters. See also **temporary hardness.**

ALKALIS are substances which **dissociate** and release OH^- ions.

ANIONS are negatively charged **ions.**

ANION EXCHANGE Exchanging the anions in a water, usually for OH^- ions.

ANION EXCHANGE RESINS Strong base resins are classed as **Type I and Type II,** both of which will **salt split.** The latter are less strongly basic and give higher **leakages** but better regenerant utilization. Weak base resins will not salt split but give very high regenerant utilisation.

ANTHRACITE can be used as a filter medium instead of sand, either for its lower density (see **Depth Filtration**) or to avoid silica being leached into the water, e.g. after **hot process.**

AS SUCH; AS CaCO₃ An analysis can show the actual concentration (usually in **ppm**) of a substance, when it is said to be 'as such'. Alternatively, the concentration of each **ion** can be divided by its **equivalent weight** and multiplied by 50, which is the **EW** of calcium carbonate. This has the advantage of showing all ions on the same equivalent basis.

ATOMS are the smallest units in which free elements can exist. They are electrically neutral, but some can take on $^+$ve or $^-$ve charges and become **ions.** The properties of such ions are quite different from those of the element.

ATOMIC WEIGHT is the weight of an atom, and the unit of measurement is the weight of the **H^+ ion.** For example, oxygen (O) has an atomic weight of 16, and therefore an atom of O weighs the same as 16 H^+ ions.

BACKWASH is carried out by lifting and loosening a bed of sand or resin by upflow, with the bed free to rise into the **rising space.** It re-classifies the bed and can be used to wash out dirt and **fines.**

BASE EXCHANGE An **ion exchange softening** process in which the **hardness** forming **ions** Ca^{++} (and Mg^{++}) are exchanged for Na^+.

BEAD Modern ion exchangers are made in the form of little spheres which are called beads.

BED see **column.**

BED VOLUME (BV) is the gross volume of the ion exchange bed. It is a very useful concept in calculating ion exchange performance.

BICARBONATES are salts containing the anion HCO_3^-. When acid is added, this ion breaks into H_2O and CO_2, and acts as a **buffer.** Also called **alkalinity.**

BOD, BIOLOGICAL OXYGEN DEMAND is an analytical test for the amount of oxygen which is used by bacteria in the course of consuming the organic impurities in water. It gives a direct measure of the degree to which an effluent would deplete a water course of life-giving oxygen. Numerically quite different from the results of the **COD** test.

BODY FEED, BODY DOSING Continuously or intermittently dosing **filter medium** or powdered activated carbon into water upstream of a **filter.**

BRACKISH WATER is above the salinity allowed for potable water (500**ppm**) but less saline than sea water (30 000**ppm**).

BREAKPOINT CHLORINATION means chlorinating water until the **chlorine demand** is satisfied and any further chlorine added remains as free available chlorine.

BRINE RECOVERY A flow sheet sometimes used for **base exchange** in which some of the spent brine is retained and re-used to achieve a lower overall salt consumption.

BRINE WASHING is used to restore the effectiveness of anion resins which have been **fouled** by **organic matter.**

BUFFER A chemical such as **bicarbonate** which changes its form when acid or alkali is added, and minimizes the change in **pH** which would otherwise take place.

BURIED DISTRIBUTORS are collectors a short distance below the surface of an ion exchange bed which feature in some **counterflow** designs. Not to be confused with **centre distributor.**

CaCO₃ Calcium carbonate is used as a standard for calculating equivalent chemical concentrations. It has an **equivalent weight** of 50.

CANDLE FILTER see **Precoat Filter.**

CAPACITY The total capacity of an **ion exchanger** is the sum of all its **active groups,** expressed in such units as **g CaCO₃/l,** or **g equiv/l.** Sometimes it is measured in g equiv/g (dry weight) but this is more for academic use. The word 'capacity' usually means the working capacity, which represents that part of the total capacity which is in practice being used for ion exchange; it depends on the regeneration level and other factors, but is often about one third of the total capacity.

CARBON FILTERS using activated carbon are used to adsorb **organic matter** or chlorine from water.

CARBONACEOUS ION EXCHANGERS are largely obsolete. They were made by sulphonating coal, yielding mixed strongly and weakly acidic exchangers. See **Cation Exchangers.**

CAPILLARY MEMBRANES (or Hollow Fine Fibre, or HFF) are **reverse osmosis** membranes about the thickness of a human hair.

CARBOXYLIC refers to the weakly acid group — COOH, which is the **active group** in **weakly acidic resins**

CARRYOVER from **clarifiers** is almost inevitable. Some particles (of **floc** etc) just do not settle e.g. because they are attached to minute air bubbles. These are called stragglers. At times there is malfunctioning and substantial carryover of normal floc can take place. Carryover must not be confused with **post-precipitation.**

CARTRIDGE FILTERS and **ION EXCHANGERS** are units which are discarded or replaced when exhausted. Cartridge ion exchangers may be returned to a servicing organisation who regenerate the exchange material, on an exchange basis.

CATALYSTS are substances which promote chemical change but remain unchanged themselves. Manganese oxide catalysts are used to promote **oxidation** of dissolved iron in filters.

CATIONS are positively charged **ions.** Many of them are formed from atoms of metallic elements.

CATION EXCHANGE Exchanging the **cations** in a water for H^+ ions.

CATION EXCHANGE RESINS are divided into strongly acidic (or sulphonic) and weakly acidic (or carboxylic). The words 'cation exchanger' used by themselves normally refer to strongly acidic material. Strongly acidic exchangers will **salt split** but use a large excess of regenerant, and vice versa.

CENTRE DISTRIBUTOR in a **mixed bed** is located at or near the interface between cation and anion resins after **separation.**

CFR is short for **counterflow regeneration.**

CHATILLON TEST is an accelerated life test method simulating chemical and physical attack on ion exchange resins.

CHELATING RESINS are special ion exchangers made to have specific **selectivity** for particular ions. Because this selectivity binds the ion very strongly, **elution** is often inefficient or complicated.

CHLORINE DEMAND is the amount of chlorine which the impurities (mostly **organic**) in a water will take up in chemical combination.

CHROMATOGRAPHY is the property of a **column** of resin to take up ions in bands, depending on their **affinity.**

CLARIFIERS are similar to **thickeners** but the term implies treating a water containing less impurity and producing a lighter sludge.

COAGULANT, COAGULANT AID The coagulant is normally a salt of iron or aluminium, introduced into water in conditions in which iron or aluminium hydroxide will precipitate. Coagulant aids are organic **macro-ions** whose charges help small particles to adhere together to form larger agglomerates.

COAGULATION follows **flocculation** when the small particles of floc begin to stick lightly together to form bigger flocs which then settle out more easily. **Coagulant aids** can be added to the process to promote this.

CO-CURRENT is the simplest and conventional method of **regeneration,** when the regenerant flows in the same direction as the water being treated.

COD, CHEMICAL OXYGEN DEMAND is an analytical test for the amount of organic matter in a water, obtained by subjecting it to chemical oxidation – in contrast to **BOD** which uses biological oxidation. It is mostly used for industrial effluents rich in organic matter.

COLLOIDS are particles too big to be considered to be in solution but so small that agitation by the normal movement of water molecules prevents them settling out. They are normally defined as having a size range of 0.005 to 0.2 microns.

COLOUR is usually due to dissolved **organic** matter; and is sometimes used to measure the quantity of organics present. Sometimes the organics themselves are loosely called the 'colour'. Colour removal can be achieved by **flocculation,** or by **adsorption** on ion exchangers or activated carbon.

COLUMN, UNIT or **BED** In most ion exchange processes the water flows through a bed of settled exchanger, within which **chromatography** can take place. It is easiest to think of such processes in terms of taking place in a tall thin column, though large units are often wider than they are tall.

CONCENTRATE see **Electrodialysis.**

CONDENSATE POLISHING Filtration and/or ion exchange of condensate. The high purity of the feed water, and even higher purity of desired product, lead to some very specific problems.

CONDUCTIVITY is the ability to conduct an electrical current, usually measured as micromhos or micro Siemens (μS), which is the conductivity of a centimetre cube of water measured as

$$\frac{1}{10^6} \text{ ohms}$$

CONDUCTIVITY COMPARATORS are instruments which measure the conductivity of two samples taken simultaneously by two **probes** set at different heights and compare them on a bridge circuit.

CONDUCTIVITY WATER is an alternative name for **ultra-pure water.**

CONTACT TIME refers to the time for which resin and regenerant are in contact. Because of problems with **displacement** the exact time is difficult to estimate.

CONTINUOUS ION EXCHANGE is usually a misnomer. It describes systems in which the resin is moved to different parts of the apparatus for the exhaustion, regeneration and rinse operations. This movement is usually intermittent and usually means interrupting the flow of treated water, but these interruptions are very short.

COUNTERCURRENT implies a continuous flow of two fluids in opposite directions, such as in a **degasser.** It is also loosely used to describe **counterflow.**

COUNTERFLOW operation results when a fixed bed of resin is contacted with water flowing in one direction through the **column** and then regenerated with liquid flowing in the opposite direction. This is also loosely called **countercurrent.**

COUNTER-IONS are the **ions** which are loosely associated with an **ion exchanger** but which can be exchanged. The counter-ions must be equivalent to the total charge on the exchanger and have the opposite electrical sign.

CROSS-LINKING The ion exchange resin matrix consists mostly of long chain molecules, like a bowl of spaghetti. At relatively few points these chains are fixed to one another by means of a cross-linking agent, usually DVB (Divinyl Benzene.) The proportion of cross-linking in **gel resins** controls the size and volume of the pores which take up water, and hence the **water regain.** Certain forms of chemical attack cause break-up of the cross-links, which causes the resin to swell and become soft.

CRUD is the **suspended** and **colloidal** impurity, mostly of metal oxides, generated in a boiler circuit. The word is said to stand for Corrosion Residue and Undetermined Detritus, or alternatively for something between Crap and Mud.

CYCLE The use of **ion exchangers** involves treating one solution with the exchanger, and then **eluting** off the exchanged ion with another. There may be a backwashing stage as well. The whole repetitive operation is called the cycle.

DEAERATOR In water treatment jargon this usually means an apparatus for removing dissolved oxygen by scrubbing the water with steam at **saturation**

temperature, either above or below atmospheric pressure. Normal practice is to introduce water below saturation temperature, allow the bulk of the steam to condense and heat the water, while a fraction of the steam goes on through the water and is vented to atmosphere, together with the removed gases. Vacuum deaerators need ejectors or vacuum pumps to extract the incondensible gases, after a **vent condenser.**

DEALKALIZATION means removing the **alkalinity** and its associated **hardness. Lime softening** achieves this, but the term usually refers to ion exchange using weakly acidic resins. The trade name Starvation also means the same thing. See **Cation Exchange Resins.**

DECAY (QUALITY DECAY or **CONDUCTIVITY DECAY)** When ion exchangers are allowed to stand without water passing through them, the first runnings on resumption of flow tend to show poor quality.

DE-CROSSLINKING see **Cross-Linking.**

DEGASSER In water treatment jargon this usually means a packed **tower** for removing CO_2 by air blowing. Water leaves such degassers saturated with dissolved oxygen and nitrogen.

DEIONIZATION (DI), DEMINERALIZATION The removal of ions by **cation** and **anion exchange,** which yields an almost totally demineralized product water.

DEPTH FILTRATION is a process in which water passes through granular media which arrest dirt particles by a complex process, which allows the filter to hold particles smaller than the pores through the media. See also **multimedia** and **upflow** filters.

DIALYSIS A membrane process using natural diffusion of ions and solutes, as distinct from **electrodialysis** where ions are moved under the influence of an applied emf. Used in kidney machines.

DILUATE see **Electrodialysis.**

DIRECT CONTACT HEATER, DC HEATER is a device in which water and steam are contacted, the steam condenses while heating the water and the condensate adds itself to the water.

DISC FILTER see **Precoat Filter.**

DISINFECTION is any one of many possible ways of reducing microbiological contamination. Complete disinfection (which is difficult to achieve) is called **sterilization.**

DISPLACEMENT in ion exchange units takes place when water is expelled by regenerant or vice versa, first from the **rising space** and then from the **void volume** (in downflow regeneration). The terms sweetening-on and sweetening-off are used instead when ion exchange is used in processes such as sugar refining or metal recovery.

DISSOCIATION is the process of splitting by which dissolved **electrolytes** become **ions.**

DISSOLVED AIR FLOTATION (DAF) A system of clarification whereby small air bubbles are used to float **floc.** Sometimes used in preference to settlement.

DISTRIBUTOR is a device to promote **piston flow.** 'Collector', 'spreader' and 'diffuser' are also used: the first usually has water flowing in and the last two have water flowing out. A common design of distributor has a large-bore header running diametrically across the unit, with parallel laterals issuing at right angles from it. Distributors may be equipped with nozzles or they may be wrapped with a fine mesh in order to avoid resin passing through them.

DVB, DIVINYL BENZENE see **Cross-linking.**

EFFECTIVE SIZE of sand or resin is the sieve aperture (in millimetres) which retains 90% of the sample. See also **Uniformity Coefficient.**

EFFICIENCY This can be a misleading term and should not be used without care. Ion exchange efficiency is usually the ion exchange capacity obtained as a percentage of the equivalent regenerant used. **Regeneration ratio** is a more useful way of expressing this.

ELECTRICAL CHARGE The unit charge on an **ion** is one electron. For example a Cl^- ion is in fact a Cl atom which has acquired an electron, and a Ca^{++} ion is a Ca atom which has lost two electrons. In other words, when $CaCl_2$ dissociates, the two Cl atoms each make off with an electron belonging to the Ca.

ELECTRODIALYSIS (ED) An electrochemical process using ion exchange **membranes** used for converting **brackish** water to potable strength. Usually two equal streams flow through an electrodialysis plant, the one called the diluate becoming depleted in salts until it is of the required quality. The other is recirculated for economy and takes up the rejected salt. This is called the concentrate.

ELECTROLYTES are substances which will **dissociate** in solution, strong electrolytes will dissociate completely when in dilute solution, while weak electrolytes will only dissociate to a small extent. Most common mineral salts, acids and alkalis are strong electrolytes.

ELUATE is the solution which comes off any **ion exchanger** when it is being eluted and which contains the **ion(s)** which accumulated on the exchanger during the previous **cycle.**

ELUTION is usually held to mean a carefully controlled **regeneration,** probably to recover some valuable **ion** which has been put on to the exchanger.

epm, EQUIVALENTS PER MILLION is a unit of equivalent concentration calculated as **ppm** divided by the **equivalent weight.** For water analysis it is equivalent to **meq/l.**

EQUILIBRIUM is reached when a reaction goes backwards as fast as it goes fowards, so that no reaction appears to be taking place. For example, at normal pressure air is in equilibrium with about 11ppm dissolved O_2 in cold water: this means that, under those conditions, exactly as many O_2 molecules are coming out of the water as are going from the air into solution.

EQUIVALENT MINERAL ACIDITY (EMA) is a measure of the total **anions** due to strongly dissociated acids such as Cl^-, SO_4^{--}. If the **cations** in a water are removed by **cation exchange,** the EMA is converted to **FMA.**

EQUIVALENT WEIGHT (EW) is the weight of an **ion** (calculated from **atomic weights**) divided by the number of its **electrical charges.**

EXCHANGE ZONE is the band of resin in an ion exchange **column** in which exchange is actually taking place. Upstream of the exchange zone the **exhausted** resin and raw water are in **equilibrium;** downstream of the exchange zone the regenerated resin and treated water are in equilibrium.

EXHAUSTION means in practice that the exchanger will not perform any more useful work.

EXOTHERMIC reactions release heat. The commonest of these is the dilution of sulphuric acid, which can be dangerous because of the large heat release.

356

EXTERNAL REGENERATION It is sometimes more convenient to move resins out of **mixed beds** and regenerate them in vessels provided especially for that purpose, especially in **condensate polishing.**

EXTERNAL TREATMENT Treatment of water before it enters the boiler circuit.

FAST RINSE see **Rinse.**

FILMING AMINES are **internal treatment** additives which help control **pH** and improve heat transfer. They **foul** cation exchange resins.

FILTER FOUNDATION is a layer of gravel or other heavy, coarse material which does not perform a filtering function but serves as an even support for the **media** and also to assist in liquid distribution. Also called sub-fill, or underfill.

FILTER MEDIUM In **depth filtration** this refers to sand, anthracite or garnet which are used in filtration. Pure and specially graded varieties of these materials must be used. In **precoat filters** it refers to the materials such as diatomite or cellulose with which the filter coat is prepared.

FILTRATION can be thought of as straining-out suspended matter in water. However, **depth filters** do more than merely act as strainers and their true method of operation is not fully understood.

FINES are resin particles so small that they cause excessive pressure drop and lead to bad liquid distribution. New resins as delivered (unless special grades are specified), contain a percentage of fines which must be removed, and more arise in use due to deterioration of the resin.

FLASH see **Saturation Temperature.**

FLASH MIXING TANKS in **precipitation processes** are small tanks with stirring gear in which chemicals are thoroughly mixed with the water, and in which conditions are not favourable for floc formation.

FLOC is the very light and fluffy **precipitate** which is formed by, e.g. the precipitation of iron or aluminium hydroxides.

FLOCCULATION is a process in which a fluffy **floc** is deliberately caused to **precipitate** in water in order to **adsorb colloids** and **organic** matter. Aluminium and iron salts are generally used to do this.

FLUX is the rate at which a **reverse osmosis membrane** allows water to pass through it. US gal/ft^2/day (gsfd) is a unit commonly used, though m/day (m^3/m^2/day) is simpler.

FOULING is the result of allowing material which harms the performance of an ion exchanger to remain on it. It can affect the **capacity** of the resin or its speed of reaction, or both. Common foulants are **humic and fulvic acids,** and detergents on anion exchangers; iron, **coagulant aids** or **filming amines** on cation exchangers.

FREE MINERAL ACIDITY (FMA) are the **anions** of strong acids which are not paired with **cations** and are therefore free to react (see also **EMA**). Conventionally, the FMA is the acidity which causes the **pH** to be lower than 4, which is the pH at which screened methyl orange changes colour.

FULVIC ACID see **Humic Acid.**

GEL RESINS are ion exchange resins in which the resin absorbs water, which swells it and so creates channels and pores through which ions can travel. These pores disappear if the resin is dried. See also **Macro-reticular Resins.**

g equiv/l=N=1000 meq/l

GRADING see **Effective Size** and **Uniformity Coefficient.**

GRAINS/GALLON A unit now almost entirely out of date. It is equivalent to 14.29 **ppm.**

GRAVEL is coarser than coarse sand and is used as **sub-fill** for **filter media** or ion exchange resins.

GREENSAND is a natural mineral with **cation exchange** properties, now obsolete because resins are more effective, but useful for **iron removal.**

H$^+$ IONS are different from other ions in being much smaller and more mobile. Waters containing more than a trace of H$^+$ ions are acid (see **pH**).

HARDNESS is the power to form scum with soap, or scale in boilers. It is due to salts of Ca^{++} and Mg^{++} which **precipitate** under certain conditions. See also **Temporary Hardness** and **Permanent Hardness.**

HAZEN A unit of **colour** (see BS 2690 Part 9, 1970).

HEADER see **Distributor.**

HEATER BOX is a **direct contact heater** working at atmospheric pressure and open to the air. Incondensible gases are vented to the air, so that the effect is that of a crude **deaerator.**

HEATING STEAM is that fraction of the steam going into a **deaerator** which heats the water and condenses in the process.

HEIGHT OF TRANSFER UNIT, HTU see **Transfer Unit.**

HFF see **Capillary Membrane.**

HOPPER-BOTTOM TANKS are the simplest form of **clarifier.**

HOT NAKED MIXED BED doesn't mean what you think it means, but means what it says.

HOT PROCESS see **Lime Softening.**

HUMIC ACID and **FULVIC ACID** These are names for a vast number of **organic** compounds which are the products of decay of vegetable matter such as peat. They make waters acidic, frothy, and give them **colour** varying from a light yellow to dark brown. Some of them **foul** ion exchange resins. The difference between the two classes of acids is unimportant to engineers.

HYPOTHETICAL COMBINATIONS is a fairly pointless way of presenting a water analysis. It lists the inorganic salts in the order in which they would theoretically **precipitate** when the water is evaporated off, that is as ppm **as such** of $CaCO_3$, $CaSO_4$, $MgSO_4$ and so on.

INERT RESIN is used in a variety of applications, e.g. to come between ion exchange resin and outlet distributor in certain **CFR** processes, or special grades which classify between cation and anion resin in a **mixed bed** and so improve **separation.**

INJECTION refers to the stage in which regenerant chemicals are pumped into the ion exchange unit. At the beginning of injection, the unit is full of water which must be **displaced** before the whole bed comes into contact with the regenerant, just as at the end of injection the rinse has to displace the chemicals. See also **Rinse.**

INORGANIC substances are those which do not contain the element carbon,

359

with the exception that **bicarbonate** and carbonate are usually considered inorganic, although they do contain carbon.

INTERFACE The level at which the cation and anion resin components of a **mixed bed** separate, after classification by **backwash** and settling.

INTERNAL TREATMENT Dosing chemicals into boiler circuits to control the quality of water in the boiler.

IONS are charged particles dissolved in water: they may be charged **atoms** such as Na^+ or Cl^-, or a charged group of atoms such as SO_4^{--}, or even macroions which arise from charged **macromolecules.**

ION EXCHANGE is a reaction between an **electrolyte** in solution and the **counter-ion** held on the exchanger. If the resin's **affinity** requires it, the counter-ion is exchanged for the ion of the same sign in the solution.

ION EXCHANGERS are insoluble solids which are in effect huge **ions** with a vast number of electrical charges, through which water can diffuse. They must always have loosely attached to them the **equivalent weight** of **counter-ions.** Synthetic resins are now normally used as ion exchangers.

kgrn/ft³ (KILOGRAINS PER CUBIC FOOT) is an obsolescent unit of concentration used for ion exchange capacity, the words 'as $CaCO_3$' usually being implied. It is very important to avoid confusion with kg.

IRON REMOVAL requires dissolved Fe^{++} to be **oxidized** to insoluble Fe^{+++} before filtration. Catalytic filter media are often used for this.

KMnO₄ see **Potassium Permanganate.**

LAL *Limulus amoebocyte lysate* (a material obtained as an extract from Horseshoe crabs) is used for detecting pyrogens in water. Nobody can remember such an awful name, so LAL or Limulus test is the name commonly used.

LANGELIER INDEX A formula for calculating whether a water tends to be corrosive or not.

LATERAL see **Distributor.**

LAUNDER in water technology is a device for ensuring even overflow from open-top tanks.

LEAF FILTER see **Precoat Filter.**

LEAKAGE in ion exchange is the quantity of ion(s) which a **column** has failed to exchange. The greater part of any leakage is usually due to the **equilibrium** at the bottom of the column, rather than to by-passing or mechanical leakage. Also called **slip.**

LIME SOFTENING, LIME-SODA SOFTENING are **precipitation processes** for removing most of the **temporary** and total **hardness** respectively. At around the boiling point (when some silica removal can also be obtained) the process is often called **hot process.**

MACROMOLECULES and **MACRO-IONS** are **organic molecules** and ions of **colloidal** size or bigger.

MACRO-RETICULAR, MACROPOROUS These words mean the same thing. They refer to ion exchange resins which have pores even when dry. The pores are bigger than those in **gel resins,** and less likely to become clogged by macro-ions. These resins are mechanically stronger and more resistant to **osmotic shock.**

MATRIX means a framework. In ion exchange resins the matrix is frequently of **cross-linked** polystyrene, which is itself quite inert, but has **active groups** put on it in the course of resin manufacture.

MEDIUM see **Filter Medium.**

MEDIUM BASE RESIN is an ill-defined term suggesting that the strength of an anion resin is somewhere between a Type II and a weak base resin: very often such a resin has both strong and weak **active groups.** See **Anion Exchange Resins.**

MEMBRANE A thin sheet used in separation processes. In **electrodialysis** the membrane is effectively ion exchange resin in sheet form. In **reverse osmosis** the membrane is **semi-permeable. Ultrafiltration** membranes have very small pores of controlled size.

METHYL ORANGE a chemical indicator which changes colour at pH 3.8, below which bicarbonate is held to be absent. Abbreviated as MO or just M.

meq/l, MILLIGRAM EQUIVALENT PER LITRE is a unit of equivalent concentration. As it is a weight/volume unit, it is independent of density and can be used for strong solutions and resin capacity as well as for water analysis. In water analysis it is equivalent to **epm.**

mg/l see **ppm.**

MIXED BED (MB) A unit in which **cation** and **anion exchange** proceed simultaneously as the water flows through a mixture of **cation** and **anion exchange resins.** In this way very complete **demineralization** is achieved.

MOLECULAR WEIGHT is the weight of a molecule calculated as the sum of the **atomic weights.**

MOLECULES are the smallest particles of a substance which can exist, all of whose atoms are linked together by chemical bonds. For organic salts they are extremely small: a molecule of NaCl contains only two **atoms,** but **organic** compounds can be of huge molecules (or not). An ion exchange resin bead is theoretically one molecule, and so is any Bakelite moulding.

MONOBED is a proprietary term for **Mixed Bed.**

MUDBALLING If a filter or ion exchanger is allowed to accumulate clayey or sticky impurities, these tend to agglomerate the particles until large balls are formed which are heavier than the rest of the bed and sink in it. Once formed, mud balls are almost impossible to break up by washing.

MULTIMEDIA FILTERS have two or more layers of **filter medium** of different density and grading, so that the coarse, light material lies above fine, heavy material and so increases the filter's efficiency.

mval A unit used on the continent, equivalent to **meq/l.**

NAKED MIXED BED The use of a **mixed bed** as both filter and **polisher** in **condensate polishing.**

NEUTRAL EFFLUENT Demineralization plants can be designed so that the excess acid and alkali from the cation and anion exchangers respectively are self-neutralizing.

NON-IONIC DETERGENTS can be used for cleaning where the detergent comes in contact with ion exchange resins. Being non-ionic, this class of detergents does not **foul** ion exchange resins, in contrast to **ABS.**

NON-REACTIVE SILICA is any form of silica so fine that it will pass through ion exchangers, but insufficiently soluble to be taken up on them. May be due to clay, diatoms, etc.

NON-REGENERABLE ion exchange units are used e.g. in laboratories where the scale of operations does not justify the trouble of regenerating, or in **condensate polishing** where the chemical load is extremely small. See also **Cartridge** and **Powdered Resin.**

NORMAL, N This is a unit of concentration used by analytical chemists and equivalent to 1000 **meq/l.**

NOZZLE has two meanings in water treatment;
    a. A piping branch on a tank
    b. The word is loosely used for 'strainer nozzle', which is a small element, usually of plastic and provided with fine slots, used to retain sand or resin while permitting water to pass. Nozzle plates are false bottoms in tanks which are fitted with evenly distributed nozzles, on which a sand or ion exchange bed rests.

ORGANIC substances include all compounds of carbon except **carbonates** and **bicarbonates** which are considered **inorganic** (though strictly speaking this is wrong). In water engineering, 'organic' frequently means natural or waste products of animal or vegetable matter, such as **humic acids,** sewage, etc.

ORGANIC TRAP A unit containing **scavenger resin** or activated carbon, whose sole duty is to remove **fouling** matter.

OUTAGE TIME is the total time in the cycle that a unit has to be off line for **regeneration, backwash** and **rinse.**

OXIDATION is chemical combination with oxygen or similar materials, such as chlorine. Organic matter is largely destroyed by oxidation. Soluble iron, when oxidized, becomes insoluble and **precipitates.**

OXIDIZING AGENTS are chemicals which either give up oxygen in chemical reactions, or supply an equivalent element such as chlorine to combine with **reducing agents.**

OXYGEN SCAVENGING An **internal treatment** in which reducing agents such as sodium sulphite or hydrazine are dosed into the water in order to mop up any oxygen in it.

OZONE O_3 prepared electrically on the spot and used as a disinfectant.

PACKING see **Tower.**

PALL RING see **Tower.**

PARTIAL PRESSURE If a liquid contains a dissolved gas then that pressure of the same gas which is in **equilibrium** with the solution is called the partial pressure. In a mixture of gases, the partial pressure of any one gas is the total pressure times the fraction of the gas in the mixture. This fraction is measured by volume, or (which is the same thing) by the number of molecules.

PARTS PER MILLION see **ppm.**

PERMANENT HARDNESS The **hardness** of a solution is reduced by boiling it, which **precipitates** the **temporary hardness.** The hardness which is left after this is called permanent.

pH is a convenient measure of **acidity** or **alkalinity.** It actually measures the concentration of H^+ **ions** in a solution on a logarithmic scale.

PHENOLPHTHALEIN, P A chemical indicator changing colour at pH 8.3, below which CO_3^{--} does not exist.

PISTON FLOW Ideal liquid flow through a **column** has the same velocity at every point of its cross-section, so that the liquid front moves like a piston. The bigger the column diameter, the more difficult this is to achieve.

PLATE FILTER see **Precoat Filter.**

POLARIZATION In membrane processes such as **reverse osmosis** and **electrodialysis** abnormal concentrations tend to build up in the thin film of water immediately on the membrane. Beyond a certain **flux** or current density these concentrations bring the process to a complete halt, and this phenomenon is called polarization.

POLAR SOLVENT is a solvent which causes **electrolytes** to **dissociate.** Paraffin, for instance, is not polar, acetone is slightly, and water is extremely strongly polar.

POLISHING usually describes the purification of a water which is already very pure, especially in **condensate polishing.** Sewage polishing is more commonly called **tertiary treatment.**

POLYELECTROLYTES are organic **macromolecules** with a very large

364

number of **electrical charges**. **Coagulant aids** are usually polyelectrolytes.

POLYSTYRENE is an inert plastic which forms the **matrix** of most ion exchangers.

POST-PRECIPITATION can occur after **precipitation processes** if the reaction does not go to completion in the reaction tank so that a **supersaturated solution** passes on to the next process. Very often it is the ion exchange resin on which the water precipitates e.g. alum floc.

POTASSIUM PERMANGANATE, KMnO₄ is an **oxidizing** chemical used in analysis to estimate the total amount of **organic** matter present.

POWDERED RESIN Mixtures of cation and anion resins may be used on **precoat filters** in **condensate polishing**.

ppb, PARTS PER BILLION A unit of concentration equivalent to the µg/l and equivalent to a thousandth of a **ppm.**

ppm, PARTS PER MILLION is a measure of concentration. One ppm is one unit weight of solute per million unit weights of solution. In water analysis the ppm is equivalent to mg/l, and **'as such'** or **'as CaCO₃'** should always be stated.

PRECIPITATION occurs when a chemical reaction in water throws a solid product of the reaction out of solution.

PRECIPITATION PROCESS includes **flocculation** and **lime** or **lime-soda softening**.

PRECOAT FILTER A relatively coarse aperture filter designed to retain a coat of **filter medium** on an extended surface. The medium is applied evenly before the start of a run and washed to drain when it is clogged. Different designs of filters use candles, leaves, plates or discs.

PRECOAT FILTRATION uses a fine dispersion of a filter medium such as cellulose or diatomite or **powdered resin,** which is deposited in a thin layer on a coarser carrier. When the filter pressure loss reaches a limit, the precoat is discharged and a new coat put on.

PROBES take samples of treated water from near the bottom of ion exchange beds to monitor approaching **exhaustion**. Silica probes are arranged to monitor a rise in **conductivity** which indicates that silica is about to break through the bottom of the unit.

PYROGENS are organic impurities which make water unfit for injection and similar pharmaceutical uses. **Sterile** water is not necessarily pyrogen-free.

RAKE FILTERS see **Air Scour.**

RAPID REACTOR is a **lime softener** in which the **precipitation** takes place on **seeded** particles.

RASCHIG RING see **Tower.**

REACTION ZONE in ion exchangers is sometimes called ion **exchange zone.** It is a relatively small section of the column in which exchange is actually taking place, and which travels down the **column** during the run.

RECOVERY in **reverse osmosis** is the percentage of treated water produced from the water fed to the plant.

REDUCING AGENTS are chemicals which tend to combine with oxygen, i.e. the opposite of **oxidising agents.**

REGENERATION serves to put the desired **counter-ion** back on the exchanger. This usually means displacing an ion of higher **affinity** with one of lower, and therefore calls for a change of concentration of the **electrolyte.**

REGENERATION LEVEL is loosely used to mean one of two things: it means either the amount of regenerant used in an ion exchange **cycle,** or the working **capacity** of the ion exchange cycle which follows.

REGENERATION RATIO is the ratio of chemicals used over the equivalent ion exchange **capacity** obtained. See also **Efficiency.**

REJECTION is the percentage of a given ion which a **reverse osmosis** membrane will retain while allowing water to pass. **Membranes** show different rejections for different ions.

RESINS are man-made **organic macromolecules.** In water treatment 'resin' usually stands for ion exchange resin.

RESIN TRANSFER is the movement of ion exchange resins from one vessel to another, such as in **external regeneration** and **continuous ion exchange.**

RESIN TRAP A device containing strainers to catch resin which has been accidentally backwashed out, or which has passed through with treated water due to a mechanical fault in the ion exchange unit.

RESISTIVITY is the specific resistivity of a cubic centimetre measured between two faces, used instead of **conductivity** to characterize ultra-pure waters.

REVERSE OSMOSIS (RO) A process originally developed for converting **brackish** water to potable strength. Water is forced through a **semi-permeable membrane** which **rejects** the dissolved salts and all large molecules, particles and microorganisms.

RINSE comes after **injection** and washes out spent regenerant. When rinse starts the exchange bed is full of regenerant, and this is still reacting with the resin. For maximum efficiency therefore one often starts with a slow rinse until most of the regenerant has been **displaced,** and then goes on to a fast rinse in order to save **outage time.**

RISE RATE is the vertical water velocity in that part of a **clarifier** tank from which the **floc** etc is to settle out.

RISING SPACE is generally allowed above ion exchange beds for **backwashing.** It is normally between 0.5 and 1.5 times the bed depth.

ROTATING BIOLOGICAL CONTACTOR (RBC) A machine used to achieve aerobic treatment of sewage waste.

SADDLES see **Tower.**

SALT SPLITTING is the ability of a cation or anion exchanger to split a neutral salt and thus produce a product water containing large quantities of H^+ or OH^- ions (e.g. to turn NaCl into HCl or NaOH respectively). Strongly acidic and basic resins can do this, but their weak analogues can do so only to a minute extent.

SATURATION TEMPERATURE At a given pressure this is the boiling point of water at that pressure. If the pressure on hot water is lowered below the saturation temperature, the water boils, loses a proportion of flash steam, and finds its new (lower) saturation temperature.

SCAVENGER RESINS are porous anion exchange resins made specifically for the purpose of removing **fouling** matter. They are usually regenerated with brine or caustic brine.

SCHMUTZDECKE see **Slow Sand Filter.**

SCRAPING an ion exchange bed means the removal of the top few cms of the resin bed after a careful **backwash.** It is often the only way of removing all the **fines.**

SEEDING refers to the fact that **precipitation** often occurs more easily when the water contains **suspended** particles on which precipitate can settle. Some processes benefit from deliberate addition of seeding material.

SELECTIVITY is the preference of the resin for one ion rather than another. In the **affinity** series this preference is an orderly progression, but **chelating resins** are made to have special selectivities for particular ions.

SEMI-PERMEABLE describes membranes with minute pores through which water can diffuse, but through which salts pass with difficulty or not at all.

SEPARATION in mixed beds is the stage prior to regeneration when the cation and anion resins are separated by **backwashing.**

SERIES REGENERATION, THOROUGHFARE REGENERATION If the regenerant for two units is passed through them in series, then the first receives a double dose. Both efficiency and treated water quality benefit as a result.

SERVICE WATER is usually the net useful output of a plant, allowance having been made for treated or semi-treated water which may be used for **regeneration** and **rinse.**

SERVICE WATER IN ION EXCHANGE is the water (raw, semi-treated or treated) needed for **backwash, regeneration** and **rinse.** Do not confuse with water to service, which is the net treated water produced by the plant.

SHOCK on ion exchange resins is a sudden change of environment leading to dimensional changes. Change of temperature (thermal shock) and change of concentration of solution (osmotic shock) have this effect, and should be avoided, especially on **gel resins,** as it causes physical breakdown of the particles.

SIEMENS see **Conductivity.**

SLIP see **Leakage.**

SLOUGHING Resins with low **cross-linking** (whether they are made that way or have become de-cross-linked) may release small fragments of matrix which pass into the water as **polyelectrolytes** and can cause severe **fouling** in ion exchangers downstream.

SLOW RINSE see **Rinse.**

SLOW SAND FILTERS are large shallow sand beds through which water flows far more slowly than through a **depth filter.** They combine filtration with biological purification of water, usually for public supply, though the process is no longer economical to install. The layer of microorganisms on which the activity takes place is called the Schmutzdecke.

SLUDGE BLANKET, SLUDGE RECIRCULATION Precipitation processes work better if the new precipitate is **seeded** by old precipitate. This is done by passing the water through a layer or blanket of old sludge, or pumping some old sludge back into the process.

SOFTENING is any process which reduces or removes the **hardness. Base exchange** is commoner than **lime softening.**

SOLIDS CONTACT describes means of contacting water with sludge in order to promote precipitation of new sludge. See also **Sludge Blanket.**

SOLUBILITY is the amount of solid or gas which a liquid can hold in solution. Gas solubility depends on **partial pressure** and temperature. See also **Supersaturation.**

SPIRACTOR is a trade name for a proprietary design of **rapid reactor.**

SPIRALLY WOUND reverse osmosis modules contain a flat sheet membrane and separator meshes rolled round a central manifold like a Swiss roll.

SPLIT STREAM A process of **dealkalization** and **base exchange** in which the raw water flow is split and the two parts passed through a **cation exchange** unit and a **base exchange** unit before being reunited.

STARVATION see **Dealkalization.**

STERILE water is free of living matter (Potable water is often far from sterile.) See **Disinfection.**

STRAGGLERS see **Carryover.**

STRAINER NOZZLE see **Nozzle.**

STRATIFIED BED The use of strongly and weakly acidic or basic ion exchangers in a single shell, the weak component lying in a separate layer immediately on top of the strong component.

STRONGLY ACIDIC RESIN see **Cation Exchange Resins.**

SUB-FILL see **Filter Foundation.**

SULPHONATED COAL see **Carbonaceous Ion Exchangers.**

SULPHONIC refers to the group SO_3^- which is the **active group** on **strongly acidic** ion exchangers.

SUPERSATURATION is the condition of a solution when it is holding more of a substance in solution than the solubility allows. **Seeding,** or some other small change, can bring about **precipitation** of such a solution.

SURFACE WASH is a backwash through the **buried distributor** which leaves the bulk of the bed undisturbed.

SUSPENDED MATTER includes all large particles bigger than **colloids** which are being carried in water.

SWEETENING-ON, SWEETENING-OFF see **Displacement.**

SWEETENING LAYER A layer, e.g. of **scavenger resin** which lies on top of a working exchange resin, whose duty is to remove **fouling** matter.

TEMPORARY HARDNESS in water is the calcium and magnesium bicarbonate, which precipitates on boiling, leaving the **permanent hardness** behind. In most raw waters, the total hardness content exceeds the bicarbonate, so that temporary hardness = **alkalinity** = **bicarbonate.**

TERMINAL VELOCITY is the maximum rate at which a particle falls (in air or water) given sufficient space in which to reach its terminal velocity.

TERTIARY TREATMENT has acquired the meaning of a **polishing** process for sewage after secondary (i.e. biological) treatment.

THEORETICAL PLATE see **Transfer Unit.**

THICKENER A tank in which suspended matter settles and compacts and is continuously or intermittently discharged, while clear water overflows. For heavy precipitates, thickeners are equipped with scrapers.

THIXOTROPIC materials have a relatively low viscosity when being agitated, but tend to set like a jelly if allowed to. Sludges tend to be thixotropic.

THOROUGHFARE REGENERATION see **Series Regeneration.**

TOTAL DISSOLVED SOLIDS or **TDS** is what is left if a sample of water is evaporated away by boiling, and the residue weighed. Because **organic** matter is volatilized as well, the TDS ought to be the same as the total concentration of **inorganic** salts measured **as such.** If the determination is done at temperatures well above boiling point, this breaks down the bicarbonates and reduces the TDS value accordingly. As a method of measurement, the technique is prone to be inaccurate.

TOWER is usually used to mean a packed tower designed for **countercurrent** contact between liquid and gas. The packing may be ceramic or plastic, it may consist of rings (Raschig rings) or more sophisticated packings such as Pall rings or saddles. The duty of a tower packing is to present the maximum contact area between water and gas. In water treatment, towers are most commonly used as **degassers.** See also **Transfer Unit.**

TRANSFER UNIT This is a chemical engineering concept of a theoretical contact apparatus in which e.g. gas and liquid moving in **countercurrent** are brought in equilibrium with one another. The efficiency of an actual contact apparatus is measured in terms of transfer units (NTUs). The height of a transfer unit (HTU) is the height of a **tower packing** whose performance equals a TU. A **theoretical plate** is much the same thing as a TU.

TRAY In gas/liquid contact apparatus such as **deaerators,** trays are used to obtain contact between the two phases. Trays can be perforated (sieve trays) or be fitted with bubble caps. The word 'plate' is sometimes used instead. See also **Transfer Unit.**

TURBIDITY is the property of scattering a beam of light which is passing through water. It is due to **colloidal** matter, and this material itself is sometimes loosely called 'the turbidity'. The property of scattering light can be used to measure the amount of colloid present.

TYPE I, TYPE II see **Anion Exchange Resins.**

ULTRAFILTRATION (UF) A process analogous to reverse osmosis, but removing only very large molecules and particulate matter.

ULTRAVIOLET RAYS can be used to **disinfect** water without adding any chemicals.

UNDERFILL see **Filter Foundation.**

UNIFORMITY COEFFICIENT measures the uniformity of particle size in sand or ion exchanger. It is the value obtained by dividing the sieve opening (in millimetres) which retains 40% of the sample by that which retains 90%.

UNIMIX A trade name for **air mixing** a single resin bed after **co-current** regeneration in order to reduce the **leakage.**

UNIT see **Column.**

UPFLOW FILTRATION see **Depth Filtration.**

USAB upflow sludge anaerobic bed.

VACUUM DEAERATOR see **Deaerator.**

VALENCY as a term is out of fashion with chemists. In ion exchange it means the number of electrical charges on an ion. Some ions can change valency, such as iron which appears as Fe^{++} or Fe^{+++}.

VENT CONDENSER Deaerators must lose a quantity of carrier steam in order to take away the gases removed from the water. To conserve heat and condensate and to avoid a large steam plume, this may pass through a condenser which allows only the incondensible gases to pass on. In vacuum deaerators, the vent condenser reduces the load on the ejector or vacuum pump.

VOID VOLUME means the volume between sand or resin particles, which is between 30% to 50% of the gross **bed volume.**

WASTE WATER is the total water wasted, including that used in **diluting** regeneration chemicals, **rinsing** and **backwashing.**

WATER PURIFICATION The purification of water to specified low limits of impurity.

WATER REGAIN is one test which is used to see whether a resin conforms to the manufacturer's specification. It measures the amount of water which the matrix takes up, as a percentage of the dry weight of the resin. De-**crosslinking** of gel resins increases water regain.

WATER TREATMENT Improvement of water quality by dosing chemicals, or partial removal of some impurities.

WEAK BASE RESIN see **Anion Exchange Resins.**

WEAKLY ACIDIC RESINS see **Cation Exchange Resins.**

WEDGE WIRE is a strainer device in which wedge-shaped wires present a narrow slot which widens downstream, thus reducing the danger of clogging.

WRAPPED LATERAL see **Distributor.**

ZEOLITE is an old-fashioned term for the ion exchanger — it usually refers to **cation exchangers,** as the word was coined long before the days of **anion exchangers.** Now restricted to mineral **ion exchangers.**

ZETA POTENTIAL An electrokinetic measurement which has been suggested for the control of **coagulation** processes. Its usefulness is not widely accepted.

ZWITTER IONS act as cations or as anions according to the environment in which they find themselves. In water technology they are usually organic **macromolecules.** Also called ampholyte ions.

Index

ABS 244
acids 80
activated carbon 13
activated sludge 279, 314
air composition 138
   flow in degassing 139
   hold down for CFR 123
   mixing 153
alkalinity 41, 101
alkalis 80
ammonia 308
   purge of anion resin 228
ammonium form in condensate
   polishing 227
ampoule washing 236
anaerobic
   contact process 297
   digestion 292, 293ff, 323
   filter process 297
   processes 281
analysis of water, accuracy 333
Aqua Purificata 233
asymmetrical membranes 197
attached-growth biological reactor 298
autocoagulation 14
autofiltration 50

backflush of UF membranes 200
backwash supply 61
backwash water in sand filters 59
backwashing in CFR 133
bacterial growth, nourishment for 248
bacterial pollution 274
base exchange 92ff
bed volume 87
belt press 325
bicarbonate 84, 170
biogas 294
biological and ion exchange process,
   combination of 172
biological filtration 279
biomass activity 303
biomass densities 299
bleed of mixed beds 158
blocking 193
BOD 273
BOD_5 test 307
body dosing 53
boiler water
   quality 107
   organics in 12
boiling-water reactor, polishing in 223
brackish water 183

breakpoint, chlorination 72
buffer 87
buffering 103, 334
buried collector in CFR 123
BV 87

calcium wash of anion resin 228
capacity of mixed bed 162
carboxylic exchangers 87
carboxylic groups
     in WBA units 91
     in WAC units 90
cation-mixed bed in condensate
     polishing 229
cellulose acetate 187
centre distributor 154
     level of 164
centrifuges 277
CFR 121
chemical cleaning of membranes 200
chemical degradation 231
chloramines 75
chloride limits in boilers 225
chlorination 13
chlorine 73
     and RO membranes 194
     dosing 74
chlorine dioxide 75
chromatography 96
chromotographic banding 168
classification 152
close hexagonal packing 164
co-flow design of unit 120
CO$_2$ 41, 101
     and water 99
     from atmospheric inleakage 224
     produced by bicarbonate 346
coagulant aids 15, 42
     point of addition 43
coagulation 13
     before flotation 29
     chemical variables in chemistry for
          flotation 30
     in filter beds 15
     liquid shear in 43
COD 273
colloids, size of 3

colour 11
     pollution by 272
column operation 94
commissioning boiler sets 215
condensate 213, 217
     filtration 220
     pH 224
conductivity 159, 345
     of pure water 211
consent procedure 272
contact stabilization 315
contact time, anion resins 149
conversion in RO 191
cooling 275
countercurrent 121, 140
counterflow 119
     regeneration 121
crud 216
cut-off properties 197

dairy wastes 292
de-alcoholized beer 199
de-sludging, control of 26
deaeration 135
deaerators 251
dealkalization 99
     by acid dosing 101
     and base exchange 105
     by sodium hydrogen blend process 104
     by weakly acidic resins 104
degassers 251
degassing 135ff
demineralization 107ff
denitrification, biological 167
densities of ion exchange resins 164
density ratio of water to steam 214
depletion polarization 184
desalting by ED 183
design of PSB unit 130
detergents 12
digesters 294
     economics of 302
disinfection 71, 238
dissociation 79
dissolved air flotation 31, 325
dissolved gases 190
dissolved oxygen 272

distillation 237
distributor design in CFR 126
driving force
    in degassing 139
    in membrane processes 178
drying beds 326

ED stack 183
electroflotation 33, 277
electrodialysis 178
electronics, water for 241
EMA 89
enriching solution by ED 183
equilibrium in ion exchange 93
equivalent concentration 340
equivalent mineral acidity 89
eutrophication 273

farmyard wastes 294
ferrous salts as pollutants 273
filter backwashing 56
    beds, air scouring of 62
        coagulation in 55
    breakthrough 55
    cake 50
    cleaning 52
    deep bed 54
    media 56
        size grading of 57
filter press 325
filterability 51
filters
    coagulation in 67
    distribution system in 63
    multi-media 56
filtration 49ff
    of compressible material 50
filtration, upflow 56
fixed-film reactors 279
flash tank for coagulation 29
flash vacuum deaerator 254
float tanks 31
flocculating agent 303
flocculation 16
flotation 27ff, 277
flow sheet for ultra-pure water 243

flow sheets in demineralization 145
flow velocities 245
    in mixed beds 154
fluidized bed 128
fluidized bed reactor 298
flux 187
    in UF 198
FMA 89
fouling 162
free mineral acidity 89

Gas Laws 251
gravity thickeners 325

hard water 162
heater box 258, 264
heater deaerator 254, 258
heating zone 253, 260
heavy metals 273
Henry's Law 99, 137, 251
high-pressure boiler and turbine
        generating system, flow sheet 216
high-pressure nuclear power-station 213
high-rate biofiltration 284
    digesters 295
    filtration 286
    media 289
high-rate filters 66
hold-up time 28
hollow fibre membrane 188
hopper-bottom tank 22
HTU 142
humic acid 12
humus 286
    settlement tanks 287
hydrochloric acid 329
hydrocyclones 277
hydrogen 303
hyperfiltration 196

Immedium filter 65
inert material in mixed beds 230
inert resin 130
    in CFR 124
    in mixed beds 164
inorganics in drinking water 76
interface level 156

ion exchange resin and condensate filters 222
ion exchange resin, choice for condensate polishing 230
ion exchange resins, weak and strong 85
ion exchange zone 95
ion exchangers 84
iron 13
iron oxides in boiler water 220

jar test 37ff
    kits for flotation 40

Kohlrausch 210

LAL test 235
leakage 86, 98
    in ion exchange 109
licking 158
lime 323
limiting condition in membrane process 182
liquid distribution in CFR 124
low-rate biological filter 284

maceration 319
magnetic filters 222
manganese 13
mass transfer 138
    in membrane processes 178
maturing process 285
meat-packing wastes 297
media 283
    depth 287
membrane cleaning 199
membrane processes for nitrate reduction 172
    ultrafiltration and pretreatment for 196ff
methane 299
methyl orange 100
microbiological growth 194
microstrainers 52
mineral media 284, 288
mixed beds 116, 152, 226
mixing in mixed beds 156
molar ratio 137

Net Positive Suction Head 262, 268
neutral effluent 150
neutral pH high-oxygen system 226
new resins in mixed beds 155
nitrate breakthrough 169
nitrate removal 166ff
nitrate selective resins 171
nitrates, medical aspects 173
nitrification 285
nitrifying bacteria 314
nitrogen as pollutant 273
nitrogen blanketing in ultra-pure water 242, 246
nitrogen scrubbing 256
non-reactive silica 14
nozzle plate 120
NPSH 262, 268
NTU 141
nutrition 304

once-through boilers 214
operating line 140
operating pressure of heater deaerators 261
ordered plastic media 289
organic compounds 4
    fouling 91
    matter 12
osmotic pressure 187
oxygen demand 273
ozone 13, 75, 313

packed beds 130
packed suspended bed 128
packed tower vacuum deaerator 263
partial pressure 136, 252
peracetic acid 247
percolating filter 283
pesticides 273
pH 16, 41, 82, 101, 159, 299
    in disinfection 72
    of boiler water 220
    range of in RO 194
pharmacopoeia 233
phenolphthalein 100
plastic media 284, 288

plastic packing 280, 287
plugging factor 194
points of use 243
polarization 181
  in reverse osmosis 184
  in ultrafiltration 185
polishing section in ultra-pure water
  systems 242
polyacrylamides 15
polychlorinated biphenyls 273
powdered resin 223, 227
precipitation 169, 278
precoat filters 53
  in condensate polishing 221
pressure filter 60
pressure used for RO 187
pressurized water reactors 222
pretreatment 11ff
  for membrane processes 202
price of chemicals and water 145
primary settlement 315
PSB 128
pulp and paper mills 214
PVDF 244
pyrogen-free water 206, 233
  systems 239
pyrogens, processes for removing 236

quality standards, drinking water 72

rabbit test for pyrogens 234
random media 289
recirculation
  in biological treatment 291
  in UF 199
recycling in CFR 131
regenerant dilution water 132
regeneration 97
  plant for mixed beds 155
  ratio 86
  zone 97
resin mix 161
resin removal in regeneration plant 231
resistivity change with temperature in
  ultra-pure water 210
restart of boiler sets 215

reverse osmosis 178, 186ff, 237
Reynolds number in membrane
  processes 179
rinse recirculation 132
  mixed beds 158
  volumes in CFR 127
RO membranes, ageing of 190
rotating biological contactor 280, 317
Royal Commission Standard 307

salt passage 190
sand filters in condensate polishing 221
scale 190
Schwebebett 128
screening 275
scrubber 267
scrubbing zone 253
SDI 193
sedimentation, lamellar 19
sedimentation 18ff
  tanks 276
selectivity of RO membranes 190
self-neutralizing 330
separation in mixed beds 155, 164
septic tanks 317, 319
series regeneration 150
service water 330
  for anion units 132
  for CFR 127
  volume in CFR 131
settlement tanks 276
sewage
  inorganic increase in 4
shifting 215
silica in high-pressure steam 214
  passage of 190
silt density index 193
size spectrum of contaminants 207
slip 86
slow rinse 127
sludge as effluent problem 24
sludge blanket 21
  blowdown control 24
  conditioning 324
  dewatering 325
  land application of 326

recirculation 22
removal for flotation 34
thickening 324
small treatment works 317
sodium hypochlorite 75
sodium limits in boilers 225
softening 92ff
solubility of gases 137
specific flow rate 87
in WAC units 90
in WBA units 90
spiral wound modules for RO 188
spray-and-tray deaerator 264
stabilization of sludge 321
static head 34
steam velocity 260
Stokes' law 18
storage tanks 246
Stork deaerator 268
strainer nozzles 120
sugar refining, contaminated condensate
in 214
sulphate limits in boilers 225
sulphonated polystyrene as foulant 231
suspended matter, pollution by 272
suspended solids 309

TDS 341
tees in ultra-pure circuits 246
temperature
in anaerobic digestion 299
in sedimentation 19
in WAC units 90
limits in RO 194
pollution by 272
range in heater deaerators 261
terminal velocity in sedimentation 18
tertiary treatment 307, 313
theoretical plate 141
thin film composite 194
total dissolved solids 341
total organic carbon 210, 235

tower packings 142
transfer unit 141
trickling filter 283
TU 141
tubular UF membranes 198

UASB 298
UF systems 222
ultraviolet 75
ultra-pure circuits, disinfection of 209
ultra-pure water 186
flow sheets 195
monitoring of 210
systems materials and construction 243
UF in 201
ultrafiltration 178, 197
uniformity coefficient 230
units of measurement 159, 339
unplasticized PVC 244
upflow anaerobic sludge blanket 298
filters 65

vacuum deaerators 260
van der Waal 14
vapour, minimum flow of 257
variations
in water analysis 333
in sewage flow 309
vent steam from deaerators 259
ventilation 290
volume of vapour in vacuum
degassing 255
volumetric loading rates 291

wash from buried distributor in CFR 124
wash water for microchip
manufacture 206
water, cost of 5
for injections 206, 233
hold-up in coagulation 29
weak base resins 88